PIMLICO

114

THE FOREST
PEOPLE

Colin Turnbull is now retired, having held
a number of teaching posts at American
universities. But he received his university
education and anthropological training at
Oxford, which confirmed his already
strongly humanistic feelings. For him field
work was the most important part of his
career, and although 'the field' was mainly
in East and Central Africa, it was also
India and Tibet, and (more recently)
Polynesia, where he lived for three years. In
the United States he was actively engaged
in academic research into the prison
system, and did much work on 'death row'
in a number of states, as a result of which he
emerged more strongly opposed than ever
to capital punishment, on both academic
and moral grounds. The same kind of
concern for humanity shows in much of his
writing.

He was ordained in India as a full
Buddhist monk by the Dalai Lama in 1992,
although his first contact with Tibetan
Buddhism was as far back as 1949. His
publications include *The Lonely African*, *The
Mountain People*, *Tibet* (with Thubten
Norbu), and *The Human Cycle*. He divides
his time between the monastery of which he
is a member in Dharmasala, and the
United States.

THE FOREST PEOPLE

COLIN TURNBULL

PIMLICO

PIMLICO
An imprint of Random House
20 Vauxhall Bridge Road, London SW1V 2SA

Random House Australia (Pty) Ltd
20 Alfred Street, Milsons Point, Sydney
New South Wales 2061, Australia

Random House New Zealand Ltd
18 Poland Road, Glenfield
Auckland 10, New Zealand

Random House South Africa (Pty) Ltd
PO Box 337, Bergvlei, South Africa

Random House UK Ltd Reg. No. 954009

First published by Jonathan Cape 1961
Pimlico edition, with a new introduction, 1993

3 5 7 9 10 8 6 4 2

Printed and bound in Great Britain by
Mackays of Chatham PLC, Chatham, Kent

ISBN 0-7126-5957-9

To Kenge, for whom the forest
was Mother and Father, Lover and Friend;
and who showed me something of the love that all
his people share in a world that is still
kind and good . . . and without evil

Acknowledgments

THIS book tries to convey something of the lives and feelings of a people who live in a forest world, something of their intense love for that world, and their trust in it. It is a world that will soon be gone for ever, and with it the people.

In whatever measure the book succeeds it is due to those who by their example have taught the way to understanding. More than any I must thank my parents, who first taught me the meaning of love, and Anandamai Ma, who for two years in India showed how the qualities of truth, goodness and beauty can be found wherever we care to look for them.

And then I must thank Professor Evans-Pritchard, a more austere teacher, who teaches all his students that the study of man should be approached, not necessarily without emotion, but with careful, scientific impartiality.

To my colleagues at the American Museum of Natural History, particularly Professor Harry L. Shapiro, I am grateful for their encouragement and interest, and for their constant reminder that an anthropologist can still be human as well as scientific.

There are many other friends who should be thanked for help in divers ways, not the least one who prefers to remain anonymous but who painstakingly read and re-read every page. But Patrick Putnam deserves special mention, because it was through his friendship and that of his wife and hospitality that I first came to know the forest. I only wish he had lived to write this book himself.

I must also acknowledge a debt to Dr. Paul Schebesta, who was the first to open the way to this wonderful world; my thanks for his encouragement, and my great respect for his tolerance of dissenting youth.

And finally I must acknowledge a debt of gratitude that can never be repaid to all those people of the forest who can neither read nor write, but who are infinitely wise, and who taught me something of that wisdom. I hope this book will convey a small part of their intense love for the forest, and their belief that it is better and kinder than the outside world that threatens to destroy it.

Certain of the photographs in this book have appeared previously in *Natural History*, the Journal of the American Museum of Natural History. The author gratefully acknowledges the permission of *Natural History* to re-publish these photographs here.

Contents

Maps

Introduction to the Pimlico Edition

Looking back on *The Forest People* is, in a way, not looking back at all. Both the forest and the people are as intensely alive as ever, as full of beauty, and as full of a goodness hard to find in other parts of the world. I count it as one of the luckiest things that ever happened to me that I was so broke on my way back from India, in 1951, that I decided to make the bulk of my journey overland through Africa, and managed to get myself side-tracked into the Ituri Forest, where it all began.

When *The Forest People* was first published a lot of people said that it could not possibly have been as lovely as I said, that I must have been exaggerating. I was told this so often, by friends and colleagues, that when I was ultimately able to return, after the independence of Zaïre and the subsequent fighting there, that I was half afraid that I might find I had indeed been overstating the case.

Not a bit of it. Both forest and people were, if anything, even more lovely than I had ever been able to say, and the longer I worked there, and as many times as I went back, the beauty only increased. Yet the Mbuti are no angels, and even for them the forest world is not an easy world in which to live. They *work* to make it 'easy', and do this by making of it a world of wonder in which there is equal room for hard work and for both love and play. Returning to the Ituri Forest has been for me, every time, like taking a fresh bath – I come out with renewed faith in humanity and its enormous potential for goodness.

This is no romantic imagining or invention, either on my part or on theirs. In fact they refer to their own greatest religious festivals, which for me reach the height of beauty and wonder, as 'work'. They will even grumble about the nights of sleep they are going to lose when they have to hold their sacred *molimo*, and are not always in the best of tempers when they begin their nightly song-fests. But then something happens, and they are transformed, and their ritual becomes a love-making that in our more frigid world is difficult to imagine. It is easy enough for one human being to love another, but in their *molimo* the

5

Mbuti bring the same human passion, through song, to the whole gloriously natural world in which they live.

The next day they seem to be normal human beings again, if somewhat exhausted and grouchy. But are they the same? I know I wasn't.

Luckily, on my first visit to the Ituri I was not yet an anthropologist. I was a would-be musician/philosopher with no real credentials in either. So I was able to enter Mbuti society with a verve I might never have felt otherwise. From the outset it was a highly subjective experience, and comparing the first visit with later visits, after I had received my academic training, the worth of that deep personal involvement was, to me, indisputable. There was much to be learned from a strictly objective approach as well, but I could never accept that it had to be one or the other, though most contemporary anthropology frowned on subjectivity and personal involvement of any kind. Fortunately, my training at Oxford allowed for both.

One of the predictable results of this involvement was that I quickly came to share the prejudices, likes and dislikes, of the Mbuti. This was made all the easier because I did in fact think more highly of their music than of any other music I had heard in Africa, and because I much preferred the damp cool of their forest world to the oppressive equatorial heat of the world of the village farmers who had invaded the forest only a few hundred years earlier. But then living with the Mbuti, listening to their stories about daily life and to their legends about another kind of life, I found myself sharing their dislike of the village world, and even a mild distaste for the villagers themselves. As far as the people themselves were concerned it was nothing as strong as a dislike; it was more a puzzlement that they could be so stupid as to prefer to live in those hot and dusty villages and work as hard as they do, and be so prone to all the sickness and conflicts inherent in sedentary life; rather than live as *real* people do, in a cool forest that provides all the food and shelter and clothing anyone could possibly need, for a minimum amount of work, and with a minimum cause for conflict. If anything, I felt sorry for the villagers.

Many years later this was to tell against me, when as an anthropologist I decided that it was time for me to work among the villagers. They remembered those early days, and so did I! But nothing in my earlier experience prevented me from now seeing the same world from their point of view, whether I accepted it or not. And certainly sharing many of their hardships as farmers in a tropical rain forest, knowing full well how very different it was to live *with* the forest rather than by battling *against* it , helped me to understand them all the more readily. Of course I still though they were stupid to live like that, just as they thought I was more than a little strange to prefer living in the forest with the Mbuti!

But both worlds were essentially non-violent, the people wanting nothing more than to live in peace with each other and with the world around them. And each in their own way found an answer. The big problem for the forest people, the Mbuti, was that while essentially an exceptionally non-violent people, they were hunters, and killed for a living. As they saw it, this was a kind of original sin, going back to the mythical ancestor who first killed an antelope, then ate it in an attempt to conceal the act of killing. But, they say, man had killed, and from then onwards *all* animals (including humans) were condemned to die, whereas the forest itself, the vegetable world, never died. And so their philosophy was given form, a philosophy that led them to be such gentle hunters. At the moment of the catch, as much as they look forward to the subsequent feasting, I have never seen any expression of joy, nor even of pleasure. It is as though, at that moment, they are totally without emotion. And they say, 'If only we could learn to live without killing, we might regain our immortality.' It is easy to sneer at such 'primitive' thinking, but look at the results. Each day they set out to hunt for what they need for that day only. To kill more than is absolutely necessary would be to heighten the consequences of that original sin and confirm even more firmly their own mortality. So, like many traditional hunting peoples, they are actually great conservationists, with an unbounded respect for life, in all forms.

All this was reinforced by my experience each time I went back to the Ituri. Each time I learned more, and came away more filled than ever with admiration for this world, so very different from our own, but with so much to teach us. I learned to look much more critically at our own western concept of family, which is impoverished indeed by comparison with the much wider concept of family as understood by the Mbuti. And to look more critically at our system of justice, with which we try so desperately hard to make right out of wrong. And to look with even less enthusiasm at our economic system in which we glorify individual success and wonder why we end up with a cut-throat mentality. That may seem harsh but, through living with the villagers, I also learned to understand the reason why things are the way they are – for us as well as for them. Perhaps we cannot afford to be as loving and egalitarian as the Mbuti, in our much more problematic world. But we can still learn from them, and possibly find alternative ways of resolving at least some of our problems.

I have been lucky in being able to return to the Ituri as often as I have, for the friendships made in those days are still very much alive. Alas, many of those I knew in the earliest days are now dead. Some died while I was there. But the Mbuti notion of 'family' is not confined to the living, it includes both the dead and the unborn; so friendship, like family, is an undying thing. One old friend however is still alive, and that is Teleabo Kenge. When I last saw him he was already a grandfather, but the same swaggering, bragging, infuriating and utterly lovable and caring human that he always was.

Three years or so ago, on hearing of the death of a good friend and colleague of mine – one who had also worked, lived and loved in the Ituri and who knew Kenge well – Kenge arranged for someone to write to me: '. . . Kenge seemed very willing to do this [hold a *molimo mangbo*] but . . . what he really seemed to want was confirmation that this mattered to you . . . that the ritual (like death itself) be ripe and not empty . . . it seemed like a cry from the heart: do not forget us and we will not forget you.'

And there was a recorded message that also contained a message from the villagers with whom Joe had worked, agreeing with the Mbuti that the *molimo* should be held, 'so that the spirit of Yusefu ['Joe'] shall rest quietly . . . we are all also going to ask our ancestors to protect you . . .' And Kenge summed it up: 'We have each lost something, someone, who belonged to all of us . . . and for such a death to be a good death, it has to be made to rejoice the forest.'

Those are the voices of Forest People, Mbuti *and* villager.

Perhaps Kenge too is dead now. Perhaps I will be dead by the time you read this. It does not matter. The Mbuti believe that when our bodies die something continues, both the goodness that is inside us, and the badness. But the object of *life* is to minimize that badness, so that only goodness lives on. That is no small lesson to learn, by seeing it lived.

And when I was last there, with Kenge almost as old as I was, there was yet one more lesson to be learned. During those years of fighting that plagued the Ituri Forest, following independence, the villagers had all been forced to flee into the forest for safety, effectively placing themselves under the protection of the Mbuti. And for a number of years they lived there, and came to see the forest as not being such a harsh and hostile world after all. And the Mbuti came to see the villagers as human beings, much like themselves. So when the fighting was all over and each went back to his own way of life there was a difference. There was a mutual understanding and respect that had not been there before, and each world was a little less exclusive of the other. The villagers now admitted Mbuti to their sacred *Nkumbi* ritual as equals; one Mbuti I knew, Katchelewa Bangama, even became a ritual doctor in that great village festival, which would have been unheard of in the old days. And on their side, the Mbuti allowed the villagers to participate in *their* sacred ritual, the *molimo*. That shocked me at first, it seemed almost like a sacrilege.

But then I had brought with me the prejudices of the past, and both villagers and Mbuti had learned to outgrow prejudice.

However, the outside world is closing in on the Mbuti. Slowly but surely the forest is being exploited by others, and as that happens the hunting/gathering way of life becomes less easy. On top of which the national government, which shares so many of our western values, is slightly ashamed of having such a 'primitive' population in its midst, and seeks to 'emancipate' them, leading them to leave the forest and become farmers, like the villagers. This may well be necessary, as well as right, from a government's point of view, but I doubt if the Mbuti will find much comfort in that, or accept that such 'progress' brings them any real benefit compared with the way of life they used to know.

But the change will be gradual, and just possibly the Mbuti will bring with them, to this rather ugly modern world, something of the beauty that they have known for so long, and enrich us all.

Colin Turnbull, 1993

Glossary

BaBira The major negro village tribe with which the Epulu pygmies have contact. As with most other tribes in this area, the *Ba* is a plural prefix, the tribal name being *Bira*. Singular prefix is *Mu*, and the prefix indicating the language of the people is *Ki*. Thus a *Mu*Bira speaks the *Ki*Bira language, as do all the *Ba*Bira villagers.

BaKpara The term used by pygmies for villagers . . . if they are being respectful.

BaLese The major village tribe to the east and north-east of the Epulu district.

BaMbuti General term for all pygmies of the Ituri Forest.

BaNgwana An Islamic tribe scattered throughout the area, whose unsavoury reputation acquired in the days when they aided and abetted the slave-traders has continued right up to the present day. KiNgwana has become the *lingua-franca* of the entire area.

Banja A pair of short sticks, split into many strands at one end and clapped together or on a log as rhythmic accompaniment to certain songs of the pygmies, particularly the *molimo* songs.

Bantu General term for all tribes speaking a certain type of language, characterized by the use of prefixes, as above. The Bantu peoples extend throughout eastern, southern, and central Africa.

Baraza The meeting place of a family or village. Sometimes a large veranda built on to a house, more often a separate building consisting of a roof supported by poles, but with no walls.

Dawa 'Medicine'—a word which may refer to effective herbs or equally to magical incantations or spells.

Ebamunyama The name which the pygmies gave to me, being their condensation of the story that I told explaining my name, 'Turnbull'. (A borderland Scot saved the king's life by leaping on the back of a bull and turning it away from the king by twisting its neck. The family was rewarded with the title 'Turner of the Bull', and with a crest of a bull with a broken neck.) *Ebamunyama* literally means 'His Father Killed an Animal'.

Elima   The festival celebrating the puberty of girls.

Fito   A young straight sapling, up to twelve feet in length, used in the construction of house frames, also as whips during certain ceremonies.

Itaba   A sweet root found only in the forest, considered a delicacy by the pygmies.

Kangay   The gardenia fruit, used by pygmies to extract a black juice used for staining the body or for drawing patterns on bark cloth.

Kenge   The name of an antelope, sometimes of the Okapi. Such names are occasionally, but not often, given to children.

Kumamolimo   'The heart of the molimo'—the central meeting place of the men's religious organization.

Lendu   A forest antelope, the size of a large dog, and with sharp pointed horns.

Liko   A brew made by the pygmies of forest berries, herbs, and cola nut.

Machetti   A large bladed knife used for cutting *fito*, etc.

Mangese   The pygmy term, also used by some BaBira, for the elders. But as used by pygmies it can also refer to a youngster, if he has shown himself to be a 'Great One' by some spectacular hunting achievement, say killing a buffalo or an elephant single-handed.

Mboloko   The smallest forest antelope, barely the size of a rabbit.

Mgongo   A large heart-shaped leaf, sometimes nearly two feet across, used for thatching the roofs of huts by both the pygmies and the villagers in this area.

They are laid like tiles, the upper row overlapping the lower.

Ndura As used by the BaBira it merely means the forest; as used by the pygmies it means the entire world. When Kenge saw the plains beyond the forest, with barely a tree in view, he referred to it as '*ndura*'.

Ngbengbe Two short sticks, stripped of bark and clapped together to give a hollow ringing tone. Used for rhythmic accompaniment to the honey-songs of the pygmies. Possibly in imitation of the sound of a honey axe chopping to get at the hive.

Nkula Both the cola nut and the red wood of the tree are implied by this name; the wood is scraped to give a red powder which makes a stain when mixed with water. Used for both body and bark cloth decoration.

Nkusa The vine from which the pygmies make twine for their hunting nets.

Nyama This refers to any animal, but is also used as a term of abuse.

Okapi A variety of forest giraffe, found only in this area.

Panda A gambling game played with beans which are thrown by two players on the ground. The skill consists in estimating at a glance how many beans have to be added to those on the ground to make up a multiple of four.

Sindula The water chevrotan, an aquatic antelope that is particularly vicious when trapped. It is considered the best eating in the forest.

Sondu A large forest antelope.

Tippu Tip (also Tibbu Tib, etc.). The notorious slave trader of the nineteenth century. From his headquarters on Zanzibar he sent out caravans through East and Central Africa, using tribes such as the BaNgwana to round up slaves and ivory.

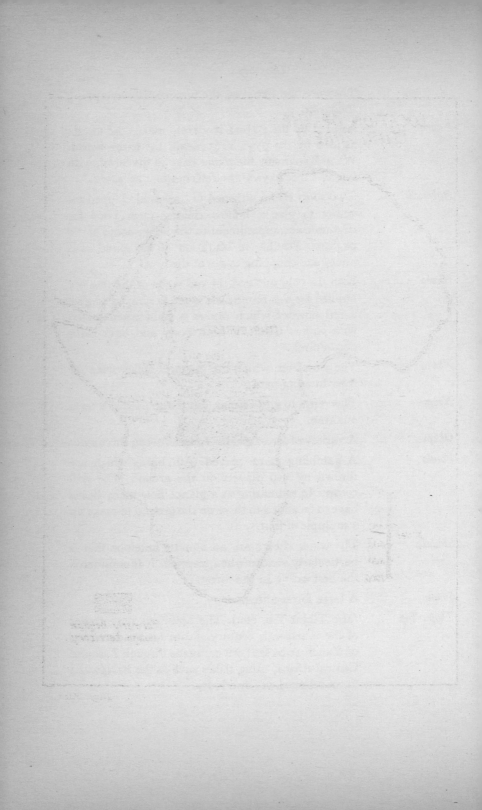

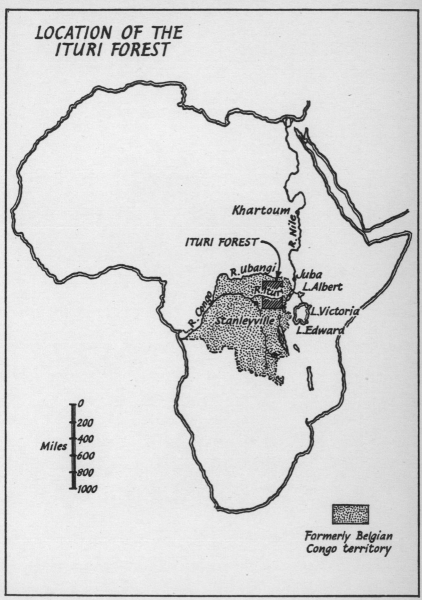

LOCATION OF THE
ITURI FOREST

Khartoum

ITURI FOREST

R. Ubangi

Juba

R. Ituri

L. Albert

R. Nile

R. Congo

Stanleyville

L. Victoria

L. Edward

Miles
0
200
400
600
800
1000

Formerly Belgian
Congo territory

Denys Baker

Chapter One

IN THE north-east corner of the Congo lies the Ituri Forest, a vast expanse of dense, damp and inhospitable-looking darkness. Put your finger right in the middle of a map of Africa and it would not be far away. Here is the heart of Stanley's 'Dark Continent', the country he loved and hated, the scene of his ill-fated expedition to relieve Emin Pasha, an expedition costing hundreds of lives and imposing almost unbearable hardships on the survivors, who trekked across the great forest not once, but three times, losing more lives each time through fighting, sickness and desertion.

Anyone who has stood in the silent emptiness of a tropical rain forest must know how Stanley and his followers felt, coming as they all did from an open country of rolling plains, of sunlight and warmth. Many people who have visited the Ituri since, and many who have lived there, feel just the same—overpowered by the heaviness of everything; the damp air, the gigantic, water-laden trees that are constantly dripping, never quite drying out between the violent storms that come with monotonous regularity, the very earth itself becoming heavy and cloying after the slightest shower. And above all such people feel overpowered by the seeming silence and the age-old remoteness and loneliness of it all.

But these are the feelings of outsiders, of those who do not belong to the forest. If you *are* of the forest it is a very different place. What seems to other people to be eternal and depressing gloom becomes a cool, restful, shady world with light filtering lazily through the tree-tops that meet high overhead and shut out the direct sunlight—sunlight that dries up the non-forest world of the outsiders and makes it hot and dusty and dirty. Even the silence felt by others is a myth. If you have ears for it the forest is full of different sounds; exciting, mysterious, mournful, joyful. The shrill trumpeting of an elephant or the sickening cough of a leopard (or the hundred and one sounds that can be mistaken for it), always make your heart beat a little unevenly, telling you that you are just the slightest bit scared, or even more. At night, in the honey season, you hear a weird, long, drawn-out, soulful cry high up in the trees. It seems to go on and on, and you wonder what kind of creature can cry for so long without taking breath. The people of the forest say it is the chameleon, telling them that there is honey nearby. Scientists will tell you that chameleons are unable to make any such sound. But in far-away Ceylon the forest people there also know the song of the chameleon.

Then in the early morning comes the pathetic cry of the pigeon, a plaintive cooing that slides from one note down to the next until it dies away in a soft, sad little moan.

There are a multitude of sounds, but most of them are as joyful as the brightly coloured birds that chase each other through the trees, singing as they go; or the chatter of the handsome black-and-white Colobus monkeys as they leap from branch to branch, watching with curiosity everything that goes on down below. And the most joyful sound of all, to me, is the sound of the voices of the forest people as they sing a lusty chorus of praise to this wonderful world of theirs—a world that gives them everything they want—a cascade of sound that echoes among the giant trees until it seems to come at you from all sides in sheer beauty and truth and goodness, full of the joy of living. And so it is, unless you are an outsider from the non-forest world, the world of human animals and savages, then I suppose this glorious song of real human beings would just be another noise to grate on your nerves.

The world of the forest is a closed, possessive world, hostile to all those who do not understand it. At first sight you might think it was hostile to all human beings, because in every village you find the same suspicion and fear of the forest, that blank, impenetrable wall. The villagers are friendly and hospitable to strangers, offering them the best of whatever food and drink they have, and always clearing out a house where the traveller can rest in comfort and safety. But these villages are set among plantations in great clearings cut from the heart of the world around them. It is from the plantations that the food comes, not from the forest, and for the villagers life is a constant battle to prevent their plantations from being overgrown.

They speak of the world beyond the plantations as being a

fearful place, full of malevolent spirits and not fit for anyone to live in except animals and BaMbuti, which is what they call the pygmies. The villagers, some Bantu and some Sudanic, keep to their plantations and seldom go into the forest unless it is absolutely necessary. For them it is a place of evil. They are outsiders.

But the BaMbuti are the real people of the forest. Whereas the other tribes are relatively recent interlopers, the pygmies have been in the forest for many thousands of years. It is their world, and in return for their affection and trust it supplies them with all their needs. They do not have to cut the forest down to build plantations, for they know how to hunt the game of the forest and gather the wild fruits that grow in abundance there, though hidden to outsiders. They know how to distinguish the innocent-looking *itaba* vine from the many others it resembles so closely, and they know how to follow it until it leads them to a cache of nutritious, sweet-tasting roots. They know the tiny sounds that tell them where the bees have hidden their honey; they recognize the kind of weather that brings a multitude of different kinds of mushrooms springing to the surface, and they know what kinds of wood and leaves often disguise this food. The exact moment when termites swarm, and when they must be caught to provide an important delicacy, is a mystery to any but the people of the forest. They know the secret language that is denied all outsiders and which makes life in the forest, for them, an impossibility.

The BaMbuti roam the forest at will, in small, isolated bands, or hunting groups. They have no fear, because for them there is no danger. For them there is little hardship, so they have no need for belief in evil spirits. For them it is a good world. The fact they average under four and a half feet in height is of no concern to them—their taller neighbours, who jeer at them for being so puny, are as clumsy as elephants, another reason why they must always remain outsiders in a world where their life may depend on their ability to run swiftly and silently. And if the pygmies are small, they are powerful and tough.

How long they have lived in the forest we do not know, though it is considered opinion that they are amongst the oldest inhabitants of that vast continent. They may well be the original inhabitants of the great tropical rain forest, stretching nearly from coast to coast. But they were certainly well established there at the very beginning of historic times.

The earliest recorded reference was not Homer's famous reference to the battle between the pygmies and the cranes, as one might think, but a record of an expedition sent from Egypt in the Fourth Dynasty, some twenty-five hundred years before the Christian era, to discover the source of the Nile. In the tomb of the Pharaoh Nefrikare is preserved the report of his commander, Herkouf, who entered a great forest to the west of the Mountains of the Moon and discovered there a people of the trees, a tiny people who sing and dance to their god, a dance such as had never been seen before. Nefrikare sent a reply ordering Herkouf to bring one of these Dancers of God back with him, giving explicit instructions as to how he should be treated and cared for, so that no harm would come to him. Unfortunately that is where the story ends, though from later Egyptian records we know that by then they had become relatively familiar with the pygmies who were evidently living, all those thousands of years back, just where they are living today, and much the same kind of life characterized, as it still is, by their dancing and singing to their god.

When Homer mentions the pygmies, with reference to their battle with the cranes, he may well be relying on information from Egyptian sources; but the element of myth is already creeping in, and in the *Iliad* (Book III) he refers to them in his description of the battle between Troy and Greece:

> '*When by their sev'ral chiefs the troops were rang'd,*
> *With noise and clamour, as a flight of birds,*
> *The men of Troy advanc'd; as when the cranes,*
> *Flying the wintry storms, send forth on high*
> *Their dissonant clamours, while o'er t' ocean stream*
> *They steer their course, and on their pinions bear*
> *Battle and death to the Pygmaean race.*'*

By Aristotle's time the western world is evidently still more inclined to treat the pygmies as legend, because Aristotle himself has to state categorically that their existence is no fable, as

* Homer, *Iliad* (tr. Derby), iii, 1–7.

20

some men believe, but the truth, and that they live in the land 'from which flows the Nile'.*

Pompeian mosaics show that, whether it was believed to be fable or no, the makers of the mosaics in fact knew just how the pygmies lived, even the kinds of huts they built in the forest. But from then until the turn of the present century, less than a hundred years ago, our knowledge of the pygmies decreased to the point where they were thought of as mythical creatures, semi-human, flying about in tree-tops, dangling by their tails, and with the power of making themselves invisible. The cartographer who drew the thirteenth-century Mappa Mundi, preserved in Hereford Cathedral, placed the pygmies accurately enough, though his representations show them as sub-human monsters, if anything.

Evidently there was still some question as to their reality up to the seventeenth century, because the English anatomist, Edward Tyson, felt obliged to publish a treatise on 'The Anatomy of a Pygmie compared with that of a Monkey, an Ape, and a Man'. He had obtained the necessary skeletons from Africa, on which he based his conclusion that the so-called 'pygmie' was, quite definitely, not human. The 'pygmie' skeleton was preserved until recently in a London museum, and it is easily seen how Tyson arrived at so firm a conclusion. The skeleton was that of a chimpanzee.

Portuguese explorers of the sixteenth and seventeenth centuries were responsible for many of the more extravagant accounts. It may well be that they actually did see pygmies near the west coast, or it may be that they saw chimpanzees and mistook them for pygmies. But it is curious that they should have thought of them as having the power to make themselves invisible, and also the power, small as they were, to kill elephants. The pygmies still kill elephants today, single handed, armed only with a short-handled spear; and they blend so well with the forest foliage that you can pass right by without seeing them. As for their having tails, it is easy enough to see how this story came into being if the pygmies seen by the Portuguese dressed as they do today, and that is more than likely. Their loin cloth is made of the bark of a tree, softened and hammered

* Aristotle, *Historia Animalium* (tr. D'Arcy Wentworth Thompson).

21

out until it is a long, slender cloth, tucked between the legs and over a belt, front and back. The women, particularly, like to have an extra long piece of cloth so that it hangs down behind, almost to the ground. They say it looks well when dancing.

Some of the accounts of nineteenth-century travellers in the Congo are no less fanciful, and it was George Schweinfurth who first made known to the world, in his book, *The Heart of Africa*, that pygmies not only existed, but were very human. He was following in the path of the Italian explorer, Miani, who a few years earlier had reached the Ituri, but had died before he could return. One of the most curious and little-known stories about the pygmies is that Miani actually sent two of them back to Italy, to the Geographic Association that had sponsored his trip. The President of the Association, Count Miniscalchi of Verona, took the two boys and educated them and contemporary newspaper reports describe them as strolling the boulevards, arm-in-arm with their Italian friends, chatting in Italian. One of them even learned to play the piano. From the present Count Miniscalchi I learned that both pygmies eventually returned to Africa where one died and the other became a saddler in the Ethiopian Army. He last heard from the latter, who must already have been an old man, just before the outbreak of World War II.

Stanley describes his meetings with the pygmies in the Ituri, but without telling us much about them, and in fact we remain ignorant of anything but the actual fact of their existence until a White Father, The Rev. Paul Schebesta, set out from Vienna in the nineteen-twenties to study them.

Schebesta's first trip was an overall survey of the forest area, in which he established the fact that this was a stronghold of the pure pygmy, as opposed to the 'pygmoid' in other parts of the equatorial belt, where there has been intermarriage with negro tribes. In subsequent trips Schebesta gathered material which showed that these Ituri pygmies, whose term for themselves, BaMbuti, he adopts, are in fact racially distinct from the negro races, Bantu and Sudanic, who live around them. This fact has so far been confirmed by later genetic studies, and though we can not be sure it seems reasonable to assume that the BaMbuti were the original inhabitants of the great tropical rain forest stretching from the west coast right across to the open savannah country of the east, on the far side of the chain of lakes that divides the Congo from East Africa.

But when I came to read Schebesta's account of the pygmies it did not ring true when compared to my own experiences on my first trip to the Ituri. For instance, in one of his first books, he says that the pygmies are not great musicians, that they do not sing anything but the simplest melodies, and beat on drums and dance wild, erotic dances. Even much later, after he had come to know the pygmies better, and had spent several years in the region, when he wrote his major work which ran into several volumes only a few pages were devoted to music, he attributed little importance to it and dismissed it as simple and undeveloped. He could not have been further from the truth.

In several other ways I felt that all was not well with Schebesta's account, particularly with his description of the relationship between pygmies and negroes. He gave the impression that the pygmies were dependent on the negroes both for food and for metal products, and that there was an unbreakable hereditary relationship where a pygmy and all his progeny were inherited by negroes, from father to son, and bound to them in a form of serfdom, not only hunting for them but also working on their plantations, cutting wood and drawing water. None of this was true of the pygmies that I knew. The one thing that did tally was that although Schebesta had not seen the *molimo* (a religious festival), from what he heard about it and about similar practices among other groups of pygmies, he felt sure that it was essentially different from the practices of neighbouring negroes, however similar they might appear to be on the surface. This certainly tallied with my own experience.

The general picture that emerged from his studies was that there were, living in the Ituri Forest, some 35,000 BaMbuti pygmies divided into three linguistic groups, speaking dialects of three major negro languages. The pygmies seemed to have lost their own language though traces remained, especially in tonal pattern. Only in the easternmost group did Schebesta feel

23

that the language had survived to any recognizable extent. These were the Efe pygmies who lived amongst the BaLese, an eastern Sudanic tribe with a not very savoury reputation of cannibalism, witchcraft and sorcery.

But in spite of this linguistic difference, and the fact that these same Efe were further different in that they did not hunt with nets, but only with bow and arrow and spear, Schebesta felt that all the BaMbuti were a single cultural unit. They tended to live in small groups of from three families upwards, moving around the forest from camp to camp, though always attached to some negro village with which they traded meat for plantation products. There was no form of chieftainship, and no mechanism for maintaining law and order, and it was difficult —from Schebesta's account—to see what prevented these isolated groups from falling into complete chaos. The most powerful unifying factor, it appeared, was the domination of the pygmies by the negroes, and Schebesta cited the *nkumbi* initiation as an example of how the negroes forced the BaMbuti to accept their authority and that of their tribal lore. Remembering what I had seen, living in an initiation camp, I could not accept this point of view at all. Yet it was one shared by others, some of whom had lived in the area for years.

The explanation was simple enough, and it was not that either one of us was right and the other wrong. Whereas Father Schebesta had always had to work through negroes, and largely in negro villages, I had been fortunate in being able to make direct contact with the pygmies, and in fact had spent most of my time with them away from negro influence. And all that any other Europeans had seen of the pygmies had been either in negro villages or on negro plantations. I had seen enough of them both in the forest and in the village to know that they were completely different people in each set of circumstances. All that we knew of them to date had been based on observations made either in the villages, or in the presence of negroes.

*　　　*　　　*

Whereas my first visit to the Ituri, in 1951, had been made mainly out of curiosity, I had seen enough to make me want to return to this area for more intensive study. An ideal location was provided by a curious establishment set up on the banks of the Epulu River, in the nineteen-twenties, by an American anthropologist, Patrick Putnam.* He had gone there to do his field work, but had liked the place and the people so much that he decided to stay. He built himself a huge mud mansion, and gradually a village grew up around him and became known as 'Camp Putnam'. The pygmies treated it just as they treated any other Bantu village (the main negro tribes nearby were the BaBira and BaNdaka, with a few Muslim BaNgwana), and used to visit it to trade their meat for plantation products. This was where I first met them.

But on my second visit, in 1954, I was provided with a real opportunity for studying the relationship between the pygmies and their village neighbours. The event was the decision of the local negro chief to hold a tribal *nkumbi* initiation festival. This is a festival in which all boys between the ages of about nine and twelve are circumcized and then set apart, kept in an initiation camp where they are taught all the secrets of tribal lore, to emerge after two or three months with the privileges and responsibilities of adult status.

The *nkumbi* is a village custom, but the pygmies always send their children to be initiated together with the negro boys, at least in those areas where the practice prevails. This has been cited as an example of their dependence on the negroes and of their lack of an indigenous culture. The negroes take all the leading roles in the festival, and as no pygmy belongs to the tribes none can become a ritual specialist, so the pygmy boys always have to depend on the negroes for admission to an initiation, and for the subsequent instruction. An uninitiated male, pygmy or negro, young or old, is considered as a child—only half a man.

Only relatives of the initiated boys are allowed to live in the camp, though any adult initiated male can visit the camp during the day-time.

But it so happened that on this occasion there were no negro

* Patrick Putnam first went to the Congo in 1927 to do field work for Harvard University. Apart from one or two brief return visits to the United States he remained there until his death at the end of 1953. At Camp Putnam he established a dispensary and a leper colony, turning his home into a guest-house to help pay the expenses of his hospital work.

boys of the right age, so the only men allowed to live in the camp and stay there all night were pygmies. To go against this custom would bring death and disaster. In spite of this the negroes went ahead with the festival because it has to be held to avoid offending the tribal ancestors. They would have liked to have stayed in the camp all night, as normally instruction goes on even then, the boys being allowed to sleep only for short periods. But custom was too strong, and they had to rely on the pygmy fathers to maintain order in the camp after dark, and not to allow the children to have too much sleep.

The pygmies, however, felt bound by no such custom, as it was not theirs anyway, and they invited me to stay with them, knowing perfectly well that if I did I would bring with me plenty of tobacco, palm wine, and other luxuries. I was, after all, they said, father of all the children, so I was entitled to stay. The negroes protested, but there was nothing they could do. They felt on the one hand that I would be punished by their supernatural sanctions for my offence, and on the other they hoped to profit themselves by my presence. At least I could be expected to share in the expenses that otherwise they would have to bear of initiating the eight pygmy boys.

And so I entered the initiation camp, and saw it through from beginning to end. It was not a particularly comfortable time as we got very little sleep. The pygmy fathers were not in the least interested in staying awake simply to keep their children awake and teach them nonsensical songs, so the negroes used to make periodic raids during the night, shouting and yelling, and lashing out with whips made of thorny branches, to wake everyone up. On top of that the camp was not very well built and the heavy rains used to soak the ground we slept on; only the boys were dry, sleeping on their rough bed made of split logs. In the end we all used to climb up there and sit, because there was no room for everyone to lie down, cold and miserable, waiting for the dawn to bring another daily round of exhausting singing and dancing.

But at the end of it all I knew something about the pygmies, and they knew something about me, and a bond had been made between us by all the discomforts we had shared together as well as by all the fun. And when the initiation was over and we were off in the forest I learned still more. It was then that I knew for sure that what little had been written about the pygmies to date gave just about as false a picture as did the thirteenth-century cartographer who painted them as one-legged troglodytes. For in the village, or in the presence of even a single negro or European, the pygmies behave in a certain way. They are submissive, almost servile, and appear to have no culture of their own. But at night in the initiation camp, when the last negro had left, and way off in the forest, those same pygmies were different people. They cast off one way of life and took on another, and from the little I saw of their forest life it was just about as full and satisfactory as village life was empty and meaningless.

The pygmies are no more perfect than any other people, and life, though kind to them, is not without its hardships. But there was something about the relationship between these simple, unaffected people and their forest home that was captivating. And when the time finally came that I had to leave once more, even though we were camped back near the village the pygmies gathered around their fire on the eve of my departure and sang their forest songs for me; and for the first time I heard the voice of the *molimo*. Then I was sure that I could never rest until I had come out again, free of any obligations to stay in the village, free of any limitations of time, free simply to live and roam the forest with the BaMbuti, its people; and free to let them teach me in their own time what it was that made their life so different from that of other people.

The evening before I left, before the singing started, three of the great hunters took me off into the forest and said that they wanted to be sure that I would come back again, so they thought they would make me 'of the forest'. There was Njobo, the killer of elephants; his close friend and distant relative, Kolongo, and Moke, an elderly pygmy who never raised his voice, and to whom everyone listened with respect. Kolongo held my head and Njobo casually took a rusty arrow blade and cut tiny but deep vertical slits in the centre of my forehead and above each eye. He then gouged out a little flesh from each slit, and asked Kolongo for the medicine to put in. But Kolongo had forgotten to bring it, so while I sat on a log, not feeling very bright, Kolongo ambled off to get the medicine, and Moke wandered around cheerfully humming to himself, looking for something to eat. It began to rain, and Njobo decided that he was not going to stay and get wet, so he left, and Moke was on the point of doing the same when Kolongo returned. Obviously anxious to get the whole thing over with as little ceremony as possible and return to his warm, dry hut, he rubbed the black ash-paste hard into the cuts until it filled them and congealed the blood that still flowed. And there it still is today, ash made from the plants of the forest, a part of the forest that is a part of the flesh, carried by every self-respecting pygmy male. And as long as it is with me it will always call me back.

The women thought it a great joke when I finally got back to camp, wet and still rather shaky. They crowded around to have a look, and burst into shrieks of laughter. They said that now I was a real man with the marks of a hunter, so I would have to get married and then I would never be able to leave. Moke looked slyly at me. He had not explained the marks as having quite that significance.

It was later that evening when the men were singing that I heard the *molimo*. By then I had learned to speak the language quite well, and I had heard them discussing whether or not to bring the *molimo* out; there was some opposition on the grounds that it was 'a thing of the forest', and not of the village, but old Moke said it was good for me to hear it before I left as it would then surely not let me stay long away, but would bring me safely back.

I first heard it call out of the night from the other side of the Nepussi, where three years earlier I had helped Pat Putnam build a dam. The dam was still there, though breached by continuous flooding. The hospital where Pat had given his life lay just beyond, now an overgrown jungle, only a few, crumbling, vine-covered walls left standing, the rest lost in a wilderness of undergrowth. Somewhere over there, in the darkness, the *molimo* called; it sounded like someone singing but it was not a human voice. It was a deep, gentle, lowing sound, sometimes breaking off into a quiet falsetto, sometimes growling like a leopard. As the men sang their songs of praise to the forest the *molimo* answered them, first on this side then on that, moving

28

around so swiftly and silently that it seemed to be everywhere at once.

Then, still unseen, it was right beside me, not more than two feet away, on the other side of a small but thick wall of leaves. As it replied to the song of the men, who continued to sing as though nothing were happening, it sounded sad and wistful, and immensely beautiful. Several of the older men were sitting near me, and one of them, without even looking up, asked me if I wanted to see the *molimo*. He then continued singing as though he didn't particularly care what my reply was, but I knew that he did. I was so overcome by curiosity that I almost said 'yes'; I had been fighting hard to stop myself from trying to peer through the leaves to where it was now growling away almost angrily. But I knew that pygmy youths were not allowed to see it until they had proven themselves as hunters, as adults in pygmy eyes, and although I now carried the marks on my forehead I still felt unqualified, so I simply said no, I did not think I was ready to see it.

The *molimo* gave a great burst of song and with a wild rush swept across the camp surrounded by a dozen youths packed so tightly together that I could see nothing, and disappeared into the forest. Those left in the camp made no comment, they just kept on with their song, and after a while the voice of the *molimo*, replying to them, became fainter and fainter and was finally lost in the night and in the depths of the forest from where it had come.

It was this that convinced me that here was something that I could do that was really worth while, and that I was not doing it justice by coming armed with cameras and recording equipment, as I had on this trip. The pygmies were more than curiosities to be filmed and their music was more than a quaint sound to be put on records. They were a people who had found something in the forest that made their life more than just worth living, something that made it, with its full complement of hardships and problems and tragedies, a wonderful thing full of joy and happiness and free of care.

Chapter Two

AFTER two years' further preparation at Oxford I felt ready to return to the Ituri, to try and understand just what it was that made the people of the forest what they were, what made them so very different from the villagers all around them, what made them seem to adopt village ways with such enthusiasm one moment, and abandon them with utter unconcern the moment they left the treeless confines of the village and returned to the forest.

If you land at Lomé, you drive first of all along the coast of Togo and Dahomey, where the Atlantic surf pounds day and night on the sands, beating right up to the first line of tall, waving palm trees. Then you climb gently inland to the rolling hills of Nigeria, lightly wooded in places, but for the most part open and fertile. Crossing into the Cameroons there is an abrupt change. You pass steeply up into a massive range of mountains, through a narrow but dense belt of forest, and on still higher until you are above the tree-line. The road is appalling, covered with loose stones and boulders, the gradient as much as one-in-four in places, deeply scoured with gulleys made by the heavy rains pouring off the mountain-side.

The road twists and turns until you lose all sense of direction, but at the top there is an open, flat stretch of grassland, and at the far side of that, in the French Cameroons, you find yourself faced with a descent as uncomfortable as the climb on the British side. But here, instead of looking down on rich farmland, open and sunny, you look out over a vast sea of dense, shimmering tree-tops, so dark they are almost black. On all sides this ocean of forest stretches for as far as you can see, quiet and peaceful, asking only to be left undisturbed, or at least to be approached in peace. But the darkness of it all also carried a threat, a warning to outsiders that they would do well to stay away. The same darkness welcomes those who understand, and there is no feeling which is quite the same as the refreshing coolness of the first shadows cast by the leafy giants at the forest edge, after years spent away from its shade and shelter.

From there onwards the forest stretches eastwards and south-ward, silent and aloof, for thousands of square miles. To the west the Atlantic is only a short distance, and to the east the forest stretches right over to the Mountains of the Moon. It is still with you when you cross the Ubangi Chari river, posses-sively covering the land right down to the water's edge. On the far side of that great river there is still another thousand miles before you reach the River Aruwimi, the wide and proud tributary of the even mightier Congo. After a few hundred miles the Aruwimi becomes less wide, but more tempestuous and turbulent, and its name changes to the Ituri. This is where the forest becomes the home of the BaMbuti, right the way through to its easternmost fringe, to the north as far as it is dense enough to cast its protective shade, and for hundreds of unknown miles to the south, where it is still unexplored. This was the home of the pygmies thousands of years before Euro-peans thought of Africa itself as anything much more than a myth.

It was a wonderful feeling to be back, but the changes at Camp Putnam were so drastic that I wondered if I had done the right thing in returning. Now that Pat was dead the name of the district had been changed to 'Epulu', the name of the river that ran past his old home. And more than the name was changed. A few hundred yards on, past the turning to the old village of Camp Putnam, the road crossed the river by a rickety wooden bridge. The bridge was as rickety as ever, but the forest on either side had been cut down, and backing on to what had once been Pat's estate was an ugly, modern motel, built by an enterprising Belgian who hoped to attract tourists. The main attraction was that on the other side of the road the government had established a *Station de Chasse* for the capture of forest animals, particularly okapi, and for the training of forest elephants. Between these two innovations down by the bridge and the old entrance to Camp Putnam stretched the mud houses of the workers, with a few tiny African stores and an establishment proudly calling itself Hotel de Bière. The vil-lage of Camp Putnam remained unaltered. Even the old mud mansion still stood.

It was not only the changed appearance of Epulu that was

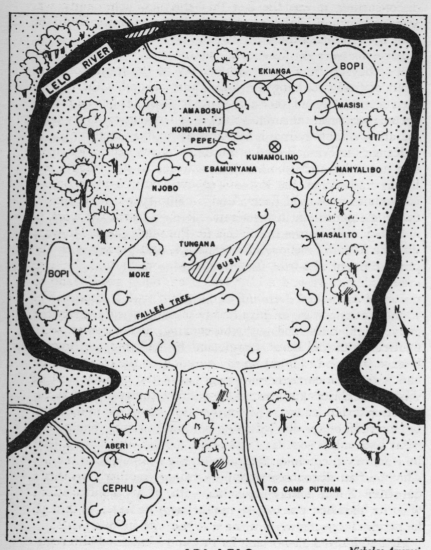

APA LELO

Nicholas Amorosi

disconcerting, it was the fact that this new community was attracting pygmy labour, particularly as pygmies could be paid even less than the other tribes. I found that several of my old friends were working either at the animal station or the motel, buying their food with money at the local stores instead of freely roaming the forest hunting and gathering for their needs. One of them, a youth called Kenge, had even become bugler at the animal station, where almost every hour the air was shattered with attempts at military bugle calls.

Kenge's father had died a number of years before, when he was just a boy. His mother had returned to her village, but Kenge had stayed. Pat Putnam encouraged him to become a jack-of-all-trades, and Kenge had come to spend more time at Camp Putnam than in the forest. He even learnt the work of hotel boy, such as pressing clothes for Putnam's guests. He wore shorts and shirt, meticulously scrubbed, and pressed with the Camp Putnam flat-iron, but usually without buttons. He was very sophisticated and a little conceited, but a good friend to me at all times, and eventually he made himself indispensable.

During my last trip Kenge had been with me practically the whole time; he helped with the cooking, as no women were allowed in the initiation camp, and he had told me how to behave. He tried to interpret my questions to the others, for he had a better idea than most pygmies of the peculiar sort of things that Europeans want to know. Such things to an African are often forbidden topics of conversation, and even to mention them would be grossly indiscreet or even insulting. Kenge became so familiar with the kind of questions I liked to ask that he would even conduct an inquiry in my absence, then tell me what he had discovered about so-and-so's mother's brother's daughter. But in leisure moments, with all his sophistication, what he liked best was to take me through the forest, as though he personally owned it, showing it to me with infinite pride.

He had insisted on working for me, and I had fired him regularly every two or three weeks because, with all his sophistication, he was as unpredictable as any pygmy, and just as disinclined to do anything he did not want to do. Several times he announced that he was going off to get food for my supper, and I would give him the money to buy something special up in the

33

village. Of course he would go straight to the beer-house and leave me to scrounge what I could for myself, and I soon learned to raid the negro plantations myself, and cut the huge banana leaves on which we used to sleep every night. On such occasions Kenge would return late, when he thought I was asleep, and lie down beside me, and very soon begin to work the blanket off my back until he was comfortably covered. If I objected he merely muttered something that I couldn't understand but that brought titters of laughter from all the other pygmies lying around the fire.

When I told him he was fired he took it as permission to take two or three days' holiday, and one morning I would wake up to find a neatly dressed and unrepentant Kenge squatting over a smoky fire concocting an especially elaborate breakfast for the two of us. In the middle of eating breakfast, without asking me if I had had enough, he often took the dish and passed it with a grand air to the other unmarried youths sitting around hungrily. If I protested he would simply tell me that I had eaten amply. If I said that I couldn't afford to feed all the bachelors in the camp he would ask me how much my Leica camera had cost. It cost me $400, and for $400 I could live in the Congo for 400 weeks. There was no effective answer. Kenge knew it, and besides, he was on his own territory, I was a mere intruder.

So when I heard that Kenge had become chief bugler for the animal station I was both relieved and sad. I knew I had been saved a lot of headaches, but Kenge was worth them all, and more. I began to look around for someone else who could take his place. Then one morning I woke up, just a few days after I had arrived, and there he was, a well-groomed pygmy in buttonless shorts and shirt, a cup of coffee balanced expertly in one hand. I asked him about his bugling, and he said he was tired of having to get up so early every morning, and anyway it hurt his lips. He announced that he was going to work for me again. I told him I could not afford to pay much as I wanted to stay a long time and had to make my money last. Kenge looked at me in the frank, open-eyed way of the real pygmy and said that he didn't mind, and named forty francs per week, a wage considerably below what he had been earning as a bugler. He said that if I could afford a present when I left, that would be fine. Pygmies have a way of letting you know when they really mean what they say, and this was one of those times. In all the many months that followed Kenge never asked me for money, though I paid him his meagre salary regularly. And once when I was short of cash and we were down in the village he spent his own savings to buy food for the two of us.

*   *   *

In the course of the previous two trips I had come to know most of the pygmies in this area quite well, and the initiation had cemented our friendship, though at the best I expect they regarded me merely as a rather harmless outsider. I was soon brought up to date on all that happened during the past three years. Kenge said that the hunting group was still hunting and gathering, but that they sometimes came down to the village and spent time there, earning money at the *Station de Chasse* or at the hotel, money which they promptly spent on food, tobacco, and palm wine at the new stores. They still brought in meat for their *BaKpara* or 'masters', and took plantation foods in return. But, encouraged by the administration, they had begun to make their own plantation in the forest. This was the worst possible news, as once the pygmies have plantations their hunting and gathering existence is made impossible. They become tied to one place and do not have the time to follow the game.

Njobo and Masisi were the two busiest workers on the plantation, and had the finest clearings. They were two of the most influential hunters of the group. Njobo had killed two elephant single-handed, and two more in company with other pygmies. He had been married three times, but only had one child, a son by his second wife. The son, Nyange, had been a healthy lad when I had last seen him, but now he was crippled with tuberculosis of the leg bone.

Masisi was a relative of Njobo, and though not such a renowned hunter he was blessed with a large family and he had a powerful and penetrating voice. He took an active part in any dispute, and usually managed to shout his opponents down, which is one of the chief ways the pygmies have of settling a dispute. His family were all strikingly fine looking, with almost Greek-like appearance although they were of normal pygmy stature, under four and a half feet. Most pygmies have unmistakable features other than height that set them apart from the negroes; their legs are short in proportion to their bodies for one; they are powerful, muscular, and usually splay backed; their heads are round and the eyes are set wide apart, with a flat nose almost as broad as the mouth is long. The head hair grows in peppercorn tufts, and the body hair varies from one extreme to the other; some pygmies are covered thickly from head to foot, a very un-negro trait. Another characteristic is the alert expression of the face, direct and unafraid, as keen as their body which is always ready to move with speed and agility at a moment's notice. These traits are all uncharacteristic of the negro tribes that live in the forest around the pygmies, a rather shifty, lazy lot who survived the ravages of Tippu Tip's slave traders by treachery and deceit. In the villages you will sometimes find pygmies that are like this too, but they are scorned by their fellow BaMbuti who have remained in the forest and have refused to be settled.

Masisi's children had finer features, with longer faces and straight noses, and they were slimmer and less stocky. His eldest son, Ageronga, was a fine hunter and often used to take his father's place.

Another member of the same clan, a cousin of Masisi, was Manyalibo. A traditionalist, he was older than Masisi, and the son of an elder brother to Masisi's father, yet he had less say in everyday matters because he was not such an active hunter. Nor did he have such a fine family. He had two daughters of his own, but no sons, and although he had adopted a distant nephew, Madyadya, the boy was proving a great trial, having been spoiled by everyone as orphans usually are. Both Manyalibo and his wife were great humorists, and made use of their powers of ridicule to break up some of the more serious disputes, because there is nothing that upsets a pygmy more than being laughed at.

But the oldest member of the group, also of the same family though even more distant, was old Tungana, who had so many children and grandchildren that neither he nor his wife could remember all their names. Strangely enough he was a progressive pygmy, and whereas Manyalibo was constantly decrying the pygmy plantation, in his deep bass voice, Tungana thought it was fine. He also thought living near the village was a good thing as there was so much good food to be had for the stealing. And wasn't stealing easier than hunting? I think the truth was that poor old Tungana was getting too old to follow the hunting band with ease, and he was afraid of being abandoned. In fact he and his wife often remained behind in the village when the rest were far off in the forest, and one of their many sons would come in every few days bringing the old couple gifts of meat and forest fruits.

Moke was another of the elders. He claimed a fictional relationship with Tungana because it made him feel at home, and he had no real relatives in the group other than his own children. Moke was the greatest traditionalist of all, and when the pygmy plantations were mentioned in front of him he just laughed dryly to himself, and smiled his toothless smile saying he had not tasted any pygmy bananas yet, but if any grew then he would be glad to eat them.

It was Moke who had come with Kolongo and Njobo to cut the marks on my forehead three years earlier. Kolongo had been as much of a traditionalist as Moke, as indeed was all his family, particularly the wizened old mother, Balekimito, and his eldest sister, Asofalinda, who was now a widow and thoroughly enjoying her freedom. Moke had thought of her as a companion for his old age, but Asofalinda liked life as it was, so Moke was still a widower and she a widow. About a year after I had left the last time, Kolongo had been killed by a crocodile, and his mother had been so upset she had left the group, her entire family, and gone back to her old village. The other son, Ekianga, remained as head of the family. It was of a different clan, but related to Njobo and Masisi and Manyalibo by marriage.

Ekianga was anything but a traditionalist in some ways, though in others he was as conservative as the rest of his relatives. His most untraditional act was to have three wives. He was a very great hunter, and as vain as could be, and he liked imitating the ways of the outside world, where it is a sign of wealth to have many wives. The pygmies say that one is enough to manage, and judging by Ekianga's constant domestic difficulties they are right. He was hairy, broad chested, and powerful almost to the point of ugliness, and from somewhere he had learned how to put on a perfect toothpaste smile. He always built his huts a different shape from everyone else, and down in the pygmy camp near the village his house was the biggest and the smartest of them all, sheltering his entire *ménage*.

His youngest wife was a beautiful girl called Kamaikan, even lighter than most pygmies, yellowish brown instead of the more usual coffee brown. Her brother and mother also lived with the same hunting group, and there was constant friction, for on marrying her Ekianga had given a 'sister', in fact a fairly distant relative, to her brother, Amabosu, in exchange. This exchange is not made without full consent of all parties, but none the less it leads to a division of loyalties, and Amabosu was a very temperamental pygmy. He was a fine hunter, but he was particularly renowned as the best singer and best drummer and best dancer in the area; by these qualities alone his prestige was enormous. His skinny old mother, Sau, was not without fame of her own. Old and infirm people, amongst the pygmies, are regarded not exactly with suspicion or mistrust, but with apprehension. In a vigorous community of this kind where mobility is essential, cripples and infirm people can be a great handicap, and may even endanger the safety of the group. So much so that there are numerous legends of how old people are left to die if they cannot keep up with the group as it moves from camp to camp. Sau knew this, as did old Tungana, and although she was still healthy she made sure everyone knew it by taking the most vigorous and unexpected part in any dispute, her sharp, acid voice betraying a certain bitterness at the way she was treated. For not only was Sau old, she was also alone, with only her son and daughter near her while everyone else had brothers, sisters, cousins, parents, children and grandchildren.

The negroes all said that Sau was a witch, and should be killed or driven away. I used to look at her, squatting over the fire outside her hut, knees hunched up to her chin, staring silently into the middle of the camp. She certainly seemed sinister, as she gazed without blinking, taking in everything that was going on, not moving for hour after hour. Her son Amabosu, the great singer, had those same staring eyes, and at times looked even more like a witch than his mother. They nearly always built their hut next to mine, and I came to be very fond of them.

If these names are confusing to the reader, they at least help to give a picture of the apparent confusion in any pygmy camp created by the complicated network of intermarriage and sister-exchange. Essentially a camp is a happy-go-lucky, friendly place, but it is also full of all sorts of little tensions that can suddenly become magnified out of all proportion and lead to a full-scale dispute. This particular group was a rather large one, consisting of the two main families, that of Njobo and Masisi, Tungana and Manyalibo, and that of Ekianga and his relatives, including Sau and Amabosu. But to add to all the little internal tensions there was a third family that was constantly trying to attach itself. It was intermarried heavily with the other two, as often happens in an attempt to strengthen bonds and establish an unbreakable relationship.

The leader of this group, although with pygmies it is always unwise to talk of single 'leaders', was a wily but rather naïve pygmy by the name of Cephu. Most pygmies amongst themselves use their real names, but they all have additional names, in KiNgwana, by which they are known to the negroes. But Cephu always used his KiNgwana name, pronounced just slightly differently from the negro equivalent, Sefu. No matter what language they are speaking, the pygmies always retain their own peculiar intonation that renders the language almost incomprehensible to non-pygmies.

Cephu had a large family, but it was not large enough, even with all his in-laws, to form a hunting group of his own. To do this you have to have at the very least six or seven individual families, each with its own hunting net; only in this way can you have an efficient net-hunt, with the women and children driving the animals into the long circle of nets, joined end to end. Cephu's group was usually not more than four families, and that was not enough, and so he tacked himself on to Njobo and Ekianga. Sometimes this worked out well enough, as pygmies are great people for visiting their relatives, and Njobo's group might be depleted in this way. But at other times they would have a number of families visiting them, and then the addition of Cephu and all his relatives made the whole group far too large and unwieldy. But as he had taken the precaution of exchanging sisters, he could not be refused, and so he would make his own little camp close by, connected by a narrow trail. He would follow the others whenever they went hunting, and was invariably blamed when the hunt was not a success. At night he and his family kept to themselves, seldom venturing into the main camp. They sat around their own fire, offended, aloof, and rather unhappy, but with hides as tough as that of a forest buffalo and impervious to the most obvious hints and thinly veiled insults. But Cephu was the best storyteller in the forest.

Kenge's position in the group was an undefined and rather happy one. His father had married a sister of Njobo, and then joined his wife's group, as often happens. But Kenge had been born to him by another wife, from a totally unrelated group. By Njobo's sister his father had one daughter, and this was Kenge's only real blood relative in the whole group. This was important, as it enabled him to flirt with almost any of the other girls, with the added spice of its being thought incestuous, because they belonged to his group, even though they were totally unrelated.

The group had changed slightly, with a few additions and subtractions, since I was last with them, and at their suggestion I decided to build my own house in their village camp. It looked as if we were in for a long spell in the village. The camp was in a strategic position; between the village of the workers of the motel and *Station de Chasse*, and that of Camp Putnam; near to a bakery, the stores, and the beer-house; and, even more important, near to all the surrounding plantations. Progressive old Tungana said that it was a good site because tourists could drive close by and take their photographs and give them money. Manyalibo, in his dry way, said the only good thing about it was that it was close to the plantations so he didn't have to go far to steal his food. The youngsters liked it because it was surrounded by the houses of negro villagers, where they could go begging for cigarettes, old clothes, and palm wine. It looked in fact as if these pygmies had become as professional as those that line the roadside at Beni, on the edge of the forest, selling their services to tourists, letting themselves be photographed doing things they would never do in the forest, wearing clothes they wouldn't wear except in a village, even selling the tourist weapons used only by negroes. But Moke reassured me. His face wrinkled up in smiles, and his soft, toothless voice said not to worry, that we would all be back in the forest long before my house was finished. 'That is where we belong,' he said, 'and we shall return soon. We can not refuse the forest.' He was right, but under circumstances he could not have foreseen, or he would not have been so cheerful.

For the first few days I went out with Kenge and others to the new 'pygmy plantation', to collect the poles and saplings for my house. The plantation was a chaotic tangle of fallen, broken, splintered wood. The pygmies, armed only with their small honey axes, and a few bigger blades borrowed or stolen from the negroes, had chopped the smaller trees down at about shoulder height, or rather less. The bigger trees they had attacked higher up, working from a flimsy frame from which they gnawed with their tiny hatchets like rats until a hundred and fifty feet of timber came crashing down. They had evidently caught on to the idea that if they attacked the big trees first, as they fell they would bring down a number of smaller ones, saving them that much work. The result was that even they found it difficult to pick their way amongst the debris, and they had no means of clearing it. None the less, hopefully, they had thrown down banana stems here and there, expecting them to take root. Njobo proudly showed me one that he had 'planted' that was struggling manfully to raise its withered leaves above the debris, into the sunshine. The only clearing had been done by negroes in search of building materials, like myself. Evidently the pygmies' enthusiasm for their plantation did not carry them very far.

The building of my house followed much the same pattern that characterizes every pygmy endeavour outside the forest world. It began with a lot of noise and big ideas. Their own camp was built in the manner of a negro village, mud huts in two lines, facing each other. The huts were only about seven feet long, and five feet deep, some of them without roofs, some without walls, none of them really complete. Larger families joined two houses together. Only Ekianga's was different, as might be expected. His house stood at the far end, and ran across between the two lines, facing down the middle. It was about twenty feet long, and it had a fine wooden door that opened and closed on hinges. The only other hut to have a door was that of Njobo, which was at the opposite end, also facing down the middle but not quite so blatantly as Ekianga's. The rest had leaf-covered frames which were simply pulled across the opening to close it.

My house was built up opposite Njobo's, with the old witch Sau and her son, Amabosu, to my right. At first the pygmies were going to build me the biggest house ever built in the forest—bigger than Camp Putnam. We finally compromised and when the framework of interlaced saplings was finished I had two rooms, each about ten foot square. Even so I knew it would never be finished, for by then the number of workers was dropping off rapidly, so I suggested just mudding the walls of one room, but leafing the whole roof so that I would have a veranda. This was thought to be a fine idea.

I was on my way out to the plantation one day, with Kenge, Manyalibo and Moke. On the way we met the handsome nephew of Cephu, a very light-skinned youth who had two children even lighter than he; his name was Kelemoke. He was wailing and crying, and told us that Cephu's daughter, only a few years old, had just died. He then continued on his way, wailing loudly. I expected the others to turn back, but they continued as though nothing had happened. Kenge made some remark about Cephu's virility and the others snorted with laughter. Kenge had the charming habit of laughing at his own jokes, which seemed funnier to him the more he thought about them, so he was still clapping his side and doubling up and shouting to the forest what a funny joke it was when we reached the plantation. Then he stopped, and although we could see nobody Moke called out quietly that Cephu's daughter had died, and Njobo and Masisi and a few others appeared from where they had been 'working', looking rather cross at being disturbed. Njobo said it was a nuisance. The girl was his wife's niece, and this meant she would keep him awake with her wailing, and would probably forget to cook for him. He set off at once for the village, grumbling, and Moke went with him. Masisi started one of his loud tirades, saying in his sharp voice that Cephu never looked after his children properly so it was no wonder they died. This must have reminded Kenge of his joke, because he started laughing again, or perhaps he was laughing at Masisi, but Masisi got angrier than ever and said he was going straight to the village to tell Cephu what he thought of him. He stamped angrily after Njobo and Moke, still shouting deprecating remarks about Cephu, the bereaved father.

Kenge laughed happily and said that there was going to be a dreadful noise when Masisi got back to the village, so we would do better to stay where we were until it was all over. Manyalibo, who had not said a word, suggested looking for *itaba*, a sweet, edible root, and with food our foremost thought we set off.

When we got back to the village in the late afternoon we found everyone sitting around outside their little houses, staring gloomily at nothing in particular. Masisi was still talking loudly, pointing at the sky and over in the direction of Cephu's camp, from where I could hear women wailing, and saying that it was a shame that Cephu should let his children die and cause everyone so much trouble. Tungana mournfully said that he didn't like funerals, the negroes made so much noise at them. Masisi's younger brother, Mambunia, still a bachelor because one of his legs was crippled with paralysis and he was not able to hunt well enough to support a family, added in shrill falsetto, that Cephu was so inconsiderate and such a trouble maker that he would probably let his daughter die completely —completely and absolutely.

I did not understand at first what he meant, but it provoked an immediate response from Asofalinda, Ekianga's widowed sister. She strode across the camp, marching with long, swinging strides to where Mambunia was sitting, his paralysed leg stuck out in front of him. She pointed her skinny hand at him, flinging her arm backwards and forwards with every step, rhythmically enumerating all the reasons why he should keep his mouth shut. He was a great one at complaining about other people being trouble makers, but he made more noise than anyone. Besides one should never say that anyone had died completely unless they had died for ever.

Madyadya, Manyalibo's adopted son, lazily wandered in between them and spitting on the ground said that Asofalinda made a pretty good noise herself, and old Moke laughed noiselessly. Asofalinda strode back to her house and slammed the fine wooden door; Mambunia continued to grumble to himself for a few minutes, and Masisi asked who was going to give him some tobacco.

It seemed that the girl was very ill with dysentery, but was not yet dead. Various degrees of illness are expressed by saying that someone is hot, with fever, ill, dead, completely or absolutely dead and, finally, dead for ever. Unhappily, early the next morning loud wailing from Cephu's camp, a quarter of a mile away, told us that the little girl was now dead for ever. When someone is ill the women relatives will wail, but it is a formal act, perhaps copied from the negroes for whom it is an obligation. But when someone really dies, for ever, there is amongst the pygmies a burst of uncontrollable grief not only amongst relatives, but among friends, women and children alike, and even men will weep if they have been close to the dead person. It is a very different sound, and a terrible one, and that is what we heard shortly after dawn.

The funeral took place the same day, directed by the negroes. I helped with the digging of the grave, but Kenge, Kelemoke, Amabosu, and Masisi's eldest son, Ageronga, were the main diggers. The negroes lent the tools and stood there, giving instructions, and they became impatient and angry when we stopped and climbed out of the grave for a smoke. The young pygmies laughed and joked, and Kelemoke who was a close relative of the girl, was in particularly good humour.

Meanwhile the body had been bathed, scented, wrapped in a white cloth and tied up in a mat placed on a rough wooden bier. This was strictly according to negro custom. In the forest the pygmies would have been unable to get the soap and scent and cloth, for one thing, and would not have had the tools to dig an elaborate grave for another. The bier was brought to the graveyard, which lay behind the village, in a noisy, wailing procession of pygmies and negroes. The body was carefully lowered into the grave. Rather grotesquely they made sure which end was which by grasping the head and twisting it under its covering. The body was oriented, laid on its side in a niche that had been cut under one wall of the grave, and held in place with three stones. Sticks were placed at an angle over it, and these were covered with leaves, then with moist earth, so that no soil should fall on the body. Then the grave was filled, some people throwing in a handful of earth, others just standing and watching. The mother and an older sister tried to throw themselves into the half-filled grave, and had to be dragged away; Cephu was weeping so bitterly that he had to be supported. Amongst the negroes there is a strict ruling about funeral protocol—who is allowed to wail and who is not, who is allowed to support the chief mourner, a purely formal gesture, and so on. But this funeral followed no such rules. As long as it made no difference to them the pygmies followed the negro custom, but the moment they wanted to go their own way they did.

Before the grave was completely covered the women left. The negroes directed the operation of guiding the spirit safely away by pouring a bucket of water, first into a hole left in the earth above the head of the corpse, then down the grave and off in the direction of the forest, away from the village. To the negroes the forest is the place for spirits of the dead, they must be kept away from the village at all costs. Rather to my surprise this struck the pygmies as being funny, and they began sniggering. Kenge, carefully dusting the last specks of earth from his spotless shorts, pretended to be severe, and said this was no time for laughter. He promptly cracked a joke at the expense of the negroes and joined his friends in howls of mirth. When pygmies laugh it is hard not to be affected; they hold on to each other as if for support, slap their sides, snap their fingers, and go through all manner of physical contortions. If something strikes them as particularly funny they will even roll on the ground, but the disapproving looks they got from the negroes discouraged them this time.

The negroes, having restored order, went on with the business of settling who was responsible for the death of the child. To them no death is natural; some evil spirit, some witch or sorcerer, had cursed the girl with dysentery and made her die. They tried to find out who had been fighting in Cephu's camp, who were the enemies of the girl's family. Their concern was a real one, for if the witch was not found, it might strike at them next. But the pygmies were not co-operative. They were bored and listless. The girl had died and that was that. The negroes finally gave it up and we all returned, washing from head to foot as we crossed over the Nepussi.

Back in the main camp everything was normal. Only in Cephu's little camp was there a quiet, subdued wailing from the house of the child's parents. There was some discussion as to whether or not the *molimo* should be called out, and Cephu was saying in an aggressive manner that it should be, the only trouble being that he didn't have one. Certain families own them, others do not, but in that case when in need they can borrow them from relatives, and Cephu promptly called on Njobo who reluctantly agreed.

That night, after the evening meal, a small number of men from the main camp went over to Cephu's camp and sat down around the central fire. The women and children were in their huts; some talking, some sleeping, some sobbing softly. The men started singing, and after a while I heard their song echoed from far off in the forest and I recognized that wistful sound, hollow and ghostly, answering the men with snatches of their own song, sometimes singing its own variations, sometimes breaking off into low, growling, animal noises. The men continued singing as though nothing were happening, and I heard the *molimo* coming closer, circling all round the camp. The women were all quiet in their huts now, and the fire had burned low; the camp was in pitch darkness except for an occasional glow where the ashes of some family hearth still shone faintly outside a hut. Cephu never built huts in imitation of the negroes, even when near a village, and his camp was a typical forest camp, a close little circle of small, conical huts, a framework of saplings covered entirely with leaves.

The *molimo* was silent for a few minutes and I looked at the faces of the men around our fire, flickering faintly in the middle of the camp. They were all staring at the flames, their eyes wide open but seeing things I had never seen. Only Cephu was lying back in his chair made of four sticks bound in the middle with a vine thong, then splayed out, and he seemed to be asleep. There was a slight movement beside me and I heard the *molimo* almost in my ear. It was a gentle sound, a little sad, now singing closely with the men, as if answering their song. Amabosu had joined the group and was on my left, and I saw he had his hands up to his mouth. Then in the shadows I saw that he was singing into what looked like a long length of bamboo, a sort of trumpet, producing that eerie, hollow sound. After a few minutes he stood up and one of Tungana's sons also stood up, holding the other end of the trumpet. The two of them danced around the fire, waving the bamboo tube over the flames, Amabosu singing into it all the time and the men singing louder and louder. Then the two performers dashed suddenly away, into the forest behind the camp, and after a few leopard-like growls the trumpet was heard no more. For about another hour the men continued singing, but as by then Cephu and the rest of his family had gone inside their huts they stopped abruptly and returned to the main camp.

The next evening the singing was around a fire in the middle of the main camp. After all, it was Njobo's *molimo*, and Cephu only went to sleep, so why should we go to his camp? A few men from Cephu's family came over, but not Cephu. The singing went on for about four hours, but the *molimo* did not appear again. The third night it was the same, only this night nobody turned up from Cephu's camp. I thought this a little strange and asked Moke about it. He said that normally one would not call out the *molimo* for a child, but Cephu had insisted; that was why nobody was really enthusiastic. Besides the *molimo* should only be held right out in the forest, not so close to a negro village. It was all right to sing *molimo* songs as long as there were no negroes about, but even then it was not good to sing them near a village. The *molimo* itself, however, should only come out in the forest, where it belonged, and there was a lot of work attached to it as it had to be given food to eat, water to drink, and fire to keep it warm. He stopped there and could not be persuaded to go any further. He just

47

waved his hands in the air and said emphatically that this *molimo* was empty, meaningless, a farce.

At the end of a week the mourning period prescribed by the negroes was over; the women who were meant to wail at sunrise, noon and sunset, stopped wailing, and everyone bustled about getting ready for the feast. A relative of Cephu's negro 'master' had come over to the village the night before, and the men had had a final burst of singing for him, but with their tongues in their cheeks. Moke told me not to think that this was the real *molimo*, it was what they did for the villagers to make the negroes think their *molimo* was the same. The villagers also make use of a musical instrument called the *molimo*, but it is meant to represent the voice of the clan totem, and it makes only animal sounds; it does not sing.

To end the period of mourning there had to be a feast, a great deal of the food being supplied by negroes. I felt that this was really why the pygmies went through with the negro funeral ceremony. Everyone was in high spirits, even Cephu and his group joined the festivities, glad that the period of mourning was over, that there was no longer any need for the women to wail, constantly reminding them of a death they would rather forget. It was better to forget the dead quickly, the pygmies said, instead of making yourself remember them all the time, as the villagers do. When referring to the Bantu in this way, disparaging them or their customs, they used one of two terms, meaning either 'animal' or 'savage'; the negroes used the same terms about the pygmies, and there was no great feeling of mutual respect.

At the height of the festivity there was a sudden cry from the roadside, swelling into a burst of wailing that was infinitely more terrible than anything I had heard before. It was certainly worse than when the little girl had died for ever. Then old Balekimito, the much-loved mother of Ekianga and his dead brother Kolongo, the mother of starchy old Asofalinda, was carried into the camp. She had been ill for some time, but being an old woman nobody had thought much of it. She had been ill before and not died completely, not even just died. But now she was completely dead, and absolutely, and her great, hulking son, the great hunter, the man of substance with three wives, hairy, ugly Ekianga, was running up and down, his face streaked with tears, beating himself on the head with his fists and crying that his mother was going to die for ever.

Balekimito was carried gently over to her son's house, through the wooden door, and laid on the floor just inside. I went in to see her. The poor old woman was so thin I could see every bone in her weak, tired body; yet when I had last seen her she had been fine, upright, strong, and a powerful old matriarch if ever there was one. She looked up at me with eyes that already had a bluish film over them, and caught hold of my hand. She said she wanted to die near her son, and was glad to be back again. She agreed to take some white man's medicine, and the dispenser from the *Station de Chasse* was sent for. This was quite a departure, as it was this family who had always resisted more strongly than any the use of modern medicine.

Ekianga was almost beside himself, and grey-haired Asofalinda, no longer stiff and rough, but full of kindness, tried to comfort her young brother, but she was too full of grief herself and knelt down on the ground beside her mother and wept. The old lady was the only one who seemed fully in control of herself, though the room was filled with wailing men, women and children. And when the dispenser came she flickered to life with a burst of her old flame, and slapped feebly at him with a bony hand as he pulled gently at her bark cloth to give her an injection, telling him to mind what he was doing. I felt her wince as the needle went in, but she continued looking up and smiling her old sweet smile. She was among her relatives and her friends, and that was all she wanted, she knew the injection would do no good. She held on to me so tightly that I couldn't leave until she finally dozed off, then when her grasp relaxed I quietly left and went out into the sunlight. Everyone was sitting around morosely, almost angrily, and even the little children whom I have seen playing contentedly through other funerals were quiet, holding on to their mothers in a frightened way. Everyone was watching Ekianga's fine wooden door, waiting for the end, waiting for old Balekimito to die for ever.

It was only a matter of hours. She never woke up.

The demonstration of grief that followed was no mere formal

expression ordained by custom, it was something very real and disturbing. I have seen death in a negro village where the atmosphere was one of fear; fear of sorcery, of the power of evil that had been unleashed. Here it was quite different, it was not a feeling of fear, but a recognition of the completeness of a loss that could never be made good. There was a finality and terrible emptiness in old Balekimito's death which could not be answered, for which there was no explanation, not even through sorcery or witchcraft. For the moment it seemed that the pygmies, faced with the death of an old and well loved and respected person such as this old lady, had nothing to cling to, and I was genuinely afraid that some of them would come to harm. Young and old alike crowded around the house, trying to force their way to Balekimito's death bed. They were even fighting to try and get in, and once inside they fought to get out. At one point several children came flying out in a frenzy, and threw themselves on to the ground, beating it with their arms and legs, kicking and biting at anyone who tried to comfort them. Inside, the commotion was even worse—Asofalinda, looking almost as old as her dead mother, had put a noose around her neck and seemed to be trying to strangle herself, full of remorse at having let her mother die this way. It took three men to take the noose from her neck, and when they finally tore it from her she ran outside and collapsed on the ground, sobbing her heart out. . . .

Even Moke quietly forced his way into the death room, and stayed there for a few minutes, saying nothing, making no sound. But tears rolled down his wrinkled face. Then he went out again and helped Asofalinda into the shade. Only Tungana and his wife, Bonyo, remained where they were, outside their hut a few yards away. They were too old for the exertion, and they just sat and cried to themselves, unashamed.

I went to my half-finished veranda to get out of the sun and found Kenge and a number of youths sitting around a fire. No matter how hot, there always has to be a fire. After a while the rest of the men joined us and sat silently, as though waiting. Then Moke came into the middle of the veranda and started talking, even more softly and quietly, so that it was difficult to hear him at all. The atmosphere was so tense I wondered if it was at last going to be a real accusation of witchcraft, though from what I had seen the pygmies had no great belief in such things. But Moke's first words put me at ease. Choking back a little sob he managed to say in a matter of fact voice, 'She died well.'

There was a general nodding of heads, and Manyalibo, who was a nephew of Balekimito, said in his gruff way that everyone should be happy that she had lived so long, happy that she had died so well, and that all this wailing should stop at once. He added that the death was a 'big thing', and that it was a 'matter for the forest'. There was a silence, and looking around to make sure that there were no women about, Manyalibo announced: 'I shall call out my *molimo*, and we will feast not for a week but for a month, or two months, or even three; we shall feast the *molimo* and make the forest happy.'

Njobo, the great elephant hunter, had the final say. He stirred the fire idly with a stick, sending showers of sparks and ashes up into the warm air. 'We have been in the village too long,' he said, 'we should have gone back to the forest before; that is where we belong. Now we must go back, far away from the village and from the people of the village. It is a bad place.'

He said that a few would stay behind to finish the work on my house, but that the rest of us should gather together all the food we could and be ready to move within a couple of days. The next day we would put Balekimito in the ground, and maybe the day after that we would leave.

At this everyone cheered up visibly and began criticizing the women and children who were still pouring in and out of Ekianga's house, beating themselves and crying. Masisi said crossly that they would never behave this way if they were in the forest, and as he talked he raised his voice until he worked himself into a real temper, at which he stalked out of the veranda and down to the far end of the little village, brandishing one hand above his head, telling everyone to stop making such a dreadful noise, that it was not doing any good to anybody. 'Crying is a matter for the immediate family,' he said. 'She was a mother to us all, but for all of us to cry is just too much noise.' He clutched his head to illustrate the point, then grabbed a couple of children, one of them his own, and sent them flying.

By now his temper was no longer artificial, and the women decided it was wiser to keep out of his way, and left.

As soon as the wailing stopped and the door to Ekianga's house closed, with only the family inside, the cloud of depression lifted. Everyone started talking about the imminent return to the forest, where they would go to hunt, where they would go for mushrooms and for fruit and nuts and honey, and where they should make their first camp. All agreed that it should be far away, but not so far that old Tungana could not come.

That night there was a little wailing from Ekianga's house, and a few women both in the main camp and in Cephu's camp were crying to themselves. Every now and then Masisi's sharp voice was raised telling them all to go to sleep.

The funeral took place the following morning, again conducted by negroes. Balekimito was buried next to Cephu's daughter, and once again when the grave had been filled in and the women had left, the negroes tried to conduct a council to discover who had been responsible for the death. But the pygmies were in no mood to play games. They simply walked off after the women, washed themselves perfunctorily in the Nepussi and returned to the camp to start making preparations for going back to the forest.

After all, there was a lot to do—plantations to raid and food to steal; and if Balekimito had died she had died well. This was no time to fool around with village notions of witchcraft and sorcery, this was a time to get back to the forest as quickly as possible and to hold the biggest *molimo* festival the forest had ever seen, to make the forest happy again.

Chapter Three

WHEN Kenge woke me the sun was not yet up, and the birds were still silent in the trees. It was three days after Balekimito's funeral, and the night before it had been decided in a general discussion that we should all leave early in the morning and build a camp at the Lelo River, several hours' walk away.

Already the pygmy village was alive with women bundling up their household possessions in the baskets they would carry on their backs. The men were busy checking their hunting nets, examining their arrow shafts and testing their bows, or sharpening the blades of their spears. The children were running about in the semi-darkness, laughing and playing, without a care in the world: only a few of the older children, around eleven or twelve years of age, helped their parents. The older girls carried an extra basket bulging with plantation foods or else they carried their baby brother or sister on one hip, bending their heads every now and again to reassure their charge with a kiss full of the warm affection of children. Some of the older boys carried their fathers' hunting nets proudly, leaving the more experienced hunters free to set off in pursuit of any game they might come across on the way.

Fires blazed in front of all the huts, and those who had not already eaten were quickly swallowing what they could before setting off. Two of Ekianga's wives, Arobanai and Loku, were already helping each other, the one supporting the basket on the other's back while the tump line around the forehead was tested for comfort and strength. In the forest, people cannot carry loads on their heads, as they do in open country, because of overhanging branches and the frequent obstacles that have to be climbed over or crawled under. So a tump line is used by the pygmies, a strip of bark a few inches wide, wound around the bundle and passed over the head. The women support this tump line by the forehead, sometimes easing the strain with their hands, but always with the head down and stooping to take more of the load on the back. The men, if they have to carry a load, generally let the tump line lie across the chest and over each shoulder, just to be different from the women.

I went over to where a group of men were drinking 'liko', a bitter brew of forest berries and nuts and herbs, bitter but warm and stimulating. They suggested that I should go on with Kenge and a couple of other youths and not wait for the rest. The women would take all day to get there anyway, they said, even though some had already left. They would stop on the way and collect mushrooms and roots for the evening meal.

Only Ekianga sat silent, outside his hut, watching his third and youngest wife, Kamaikan, cooking his breakfast. He ignored his two other wives, and they left without saying a word to him. As Kamaikan squatted over the fire, fanning it until it glowed red, I saw that she was pregnant, and I gathered from the look she cast after her two co-wives that her condition was the cause of some jealousy in the household.

Kenge delayed leaving, finding one excuse after another, until the sun was up. He then admitted that he had thought it silly to leave so early, when the forest was still wet and cold. He disappeared for an hour after that, and when he returned he was carrying a sleeping mat he had scrounged from a negro (pygmies generally sleep on leaves or sticks), a large bunch of plantains, and a cake of scented soap. He decided he could not carry my bundle as well as these things, so he told me I would have to carry my own. Anyway, he said, he wanted to leave his hands as free as possible in case he saw a bird or monkey he could shoot for dinner. He pulled the string of his bow and let it go with a loud twang, shooting an arm out in the direction the arrow would have gone, pointing to a tree-top where a black and white Colobus monkey, idly scratching its stomach, gazed down at the scene below. I heard later that Kenge considered himself above acting as a porter; he apparently thought of that as a job fit only for people of the village, not people of the forest. Any self-respecting pygmy carries his own load, and no one else's.

When the sun was above the trees and the birds were singing and the village half empty, we crossed the road and walked through the negro settlement, stopping for a quick, last drink of palm wine, for no such luxuries are to be had in the forest. There were six of us in all; five pygmy youths and myself.

The palm wine finished, we set off through the negro plantations at such a pace that I almost had to run to keep up. At the edge of the plantations the forest was thick and matted, and the path entering it was narrow. The ground was still rough where saplings had been hacked off just above ground level, and on either side was a tangle of fallen trees and ever-spreading vines and creepers. Here and there where the *fito*—the sapling used by villagers for building their houses—had once grown, the tangle was even worse. Once the forest is cut it grows up again, not as the mighty primeval forest it was, but as a jungle of secondary vegetation that in places is almost impenetrable. We went in single file, for the most part following a path, but when we left it the youths in front flashed their hatchets above their heads slashing away overhanging branches and vines without even slowing up, making it easier for us to follow. But where they cut they left sharp, jagged ends sticking out at just about eye-level for myself. When newly cut these dangerous points were plainly visible and easy to avoid, but when they became old and dirty they were a real menace.

After about half an hour we came to the place where the more ambitious pygmies had tried to make a plantation; this was the furthest point that the negroes had reached from the new village of Epulu, when they were searching for the wood to build their houses. Beyond this they did not care to venture, except for an occasional one such as Kaweki, who made his living by trapping fish in the forest streams, and those who came to help him or to carry his fish to the market. Beyond this point, in negro eyes, the forest was evil; full of animals, malevolen spirits, and pygmies.

But for the people of the forest the world only began at this point. There was a stream nearby, just beyond the 'plantation', near a site where I had recorded the forest songs of these same pygmies three years earlier. In this stream, for no obvious reason, each one of the five youths washed carefully, as though literally ridding their very feet of the dust of the village. Then they splashed though the cool, clear water and set off again, with even greater vigour than before, singing snatches of songs and shouting backwards and forwards to each other about all the animals they were going to kill and eat. Occasionally we heard an echo of our own voices coming back from the depths of the forest. Then we stopped, and one by one the youngsters shouted bold phrases to imaginary maidens, holding their noses and mouths, ready to try and stifle the shriek of laughter that would come when the same rude challenge was shouted back at them. They clapped each other on the back and held on to each other for support as they laughed, inventing all sorts of things they would do and say to any girl who answered them in such a way. The pygmy is not in the least self conscious about showing his emotions, and he likes to laugh until tears come to his eyes and he is too weak to stand. He then sits or lies down on the ground and laughs still louder.

As we went on the forest quickly became more spacious, and it seemed lighter although the trees still met high overhead and shut out all direct sunlight. There was no need to worry about overhanging branches now, for here the forest was virgin, untouched by the sacrilegious hand of the village cultivator, and it was clear of heavy undergrowth. The ground was soft underfoot, carpeted with leaves that sometimes concealed a projecting root. But the pygmies knew where every obstacle was, even though they could not see them, and they picked up their feet to avoid tripping.

On all sides we could see for twenty or thirty yards, the giant trunks of the old trees surrounded by slender, branchless saplings, all so far apart that we could wander off the path freely at almost any point. Here and there the main path was crossed by a smaller path made by some animal, and Kenge frequently turned off and followed the animal trail, beckoning me to follow. A few minutes later he would stop and sit down on his bundle, and then when the rest appeared along the main path he would jump to his feet and make fun of them for 'walking like the BaNgwana'. The BaNgwana are the most disliked of all the neighbouring negro tribes, and amongst the pygmies they are derided particularly because of their slow, ungainly gait, and as being the least able to walk in the forest as you should be able to walk; swiftly, silently, and easily.

Kenge was full of surprises. In spite of his apparent sophistication, evidenced by his attachment to western clothes and scented soap, he still knew his way about the forest better than most pygmies, and he certainly had a sharper eye for game. We stopped several times while he and another youth stalked carefully up to a tree where they had seen something. Each time they shot their arrows and stood poised on the balls of their feet, silent and motionless, only to relax and clap their hands in disgust and shake their heads as they explained to each other why they had missed. And each time all I could see of the bird or monkey was a shaking of the branches as it fled in fright.

After about two hours we came to Apa Kadi-ketu, The Camp of Kadi-ketu. Nobody was sure who or what Kadi-ketu was, but that was what the spot had always been called. This was where I had first stayed in a pygmy camp, and it was strange to pass through the same place now. There was not a sign of a hut; just the faintest outline of what had been a camp site and a slight tangle of undergrowth showing that there had been some clearing. Kadi-ketu was not quite half-way to the Lelo River, and we stopped there for a few minutes to have a smoke.

But even while resting, squatting on our bundles, Kenge and his friends were looking around, and before leaving they made a quick foraging expedition and returned in as much time as it took me to hitch up my bundle and try to find some way of fastening the tump line so that it relieved my aching shoulders. A typewriter and a load of paper and notebooks wrapped up inside a thick blanket made the load awkward and heavy. Each of my companions had handfuls of mushrooms; great fluffy pink things, fresh and firm, and tearing large leaves from the undergrowth they made small bundles and hitched them to their belts, using the stems of the leaves for twine.

Not much further on we heard loud singing ahead of us, and we came up behind the early starters who had been doing just what we had, gathering mushrooms. But they had larger bundles, and the small children were carrying them in their arms. Unlike the men the women ambled along slowly, sometimes chatting, sometimes singing. This was not simply because they were happy to be back in the forest, but because by going slowly they could keep an eye open for edible roots and fruits, and by singing and chatting loudly they scared away any of the animals that might otherwise have been startled into attack. There are few dangers in the forest, but there are plenty of buffalo and leopard that would attack without question if you came on them unawares. Even two pygmies (and they never travel singly if they can help it) will shout and clap their hands as they walk along the forest trails always in single file. The only time they are silent is when actually hunting.

We passed the women and children, and shortly afterwards we passed a second group, those who had left before dawn. They were still sitting around where they had built fires and cooked themselves a meal. Every woman, when moving camp, carries a burning ember with her, wrapped heavily in fire-resistant leaves. None of these pygmies know how to make fire. The first thing they do when they stop on the trail for a rest is to unwrap the ember, and putting some dry twigs around it they blow softly once or twice and transform it into a blazing fire. I have never learned what the knack is, and many a time I have been blowing and fanning my fire, trying to get it to come up, when a little slip of a girl, or a boy almost too young to walk, has come along, knelt down, and given two little puffs that have sent the flames leaping upwards. In the forest it is damp and cool, and at any time of the day it is pleasant to sit around the fire; but at night time when it is cold the fire not only gives warmth but it also gives protection, keeping the animals away.

It was midday when we waded through a marshy patch left by recent floods. Usually there is little marsh or swamp, but every now and again dead branches and fallen trees will block a stream and the whole area will be flooded, perhaps for miles, until a heavy rain washes the obstruction away. We had to splash knee deep through mud and water for several hundred yards before we reached dry land. The first thing Kenge and the others did then was to find the main stream and wash themselves clean before going any further.

From here the ground rose steadily and although we could hear streams below us we did not cross any more until just before we got to the Lelo. Up on the ridge the forest was a little thinner, and more light filtered through to the ground below and the world was alive with dancing shadows as the breeze rippled the leaves. But the sun still did not manage to do more than raise the temperature until it was pleasantly warm. As we were picking our way down into the Lelo valley we heard distant rumbles of thunder, and Kenge stopped and looked up into the tree-tops to see which way the wind was blowing. He muttered to himself and said that it would not rain yet, but that if those lazy women didn't hurry up and come and build the huts we would have a very wet night.

But for the moment the sun was shining, and when we finally arrived at the site for our camp it was bright and green and cheerful. It was a natural clearing a couple of hundred yards across with a patch of trees and bush in the centre, breaking it up so that it really formed two clearings. Large and as open as it was, the trees still almost met overhead. The ground, instead of being covered with leaves, was covered with a kind of grass, and Kenge dumped his bundle on top of a little hillock several yards across and said that this was where we would have our house built. He then told me to come and see the Lelo itself, hinting that I should have a good bath.

Apa Lelo, or The Camp of Lelo, will always be one of the most beautiful parts of the whole forest for me. This is partly because of what happened there, but also because it is one of the places where the forest is at its gentlest and kindest. Even when we left, after a month of heavy rains had turned the camp site into a morass of mud, it was still beautiful, and much as the pygmies hate a muddy camp they were reluctant to leave. But for the moment all was green and fresh.

Elsewhere the forest can be severe, particularly in the mountainous regions; and in many places it is overwhelming in its immensity. The leafy roof where the trees meet overhead can be so far above you that you feel as small and insignificant as the other animals roaming the ground below, depending on the forest for life. And at times it can be so powerful that it is frightening. When it whips itself into a fury, conjuring up rain and thunder and lightning, it turns small streams into raging torrents and sends heavy branches or even whole trees crashing down, some of them with trunks as thick as houses, destroying the lesser trees that stand in their way.

But even in a storm Apa Lelo was gentle and kind and protective. Here, somehow, one felt superior to the other animals of the forest. At the far end of the camp, just beyond the hillock where Kenge had said our house was to be built, a path led off to the left through a glade where the smell of leaf mould was rich and cool and moist. Termites built their nests there, elfin huts of clay with funny mushroom-shaped roofs; small animals scampered about below and birds fluttered lazily overhead. The world at Apa Lelo was full of delicate colours and shades and whispering sounds, and up in the highest branches orchids shyly hid their delicate heads amongst the moss.

This glade was almost an island, as the Lelo swept around it, all but encircling the camp. The Lelo, the river the pygmies loved above all others, widened out at the end of its curve into a broad, shallow stream, a hundred feet or so across, clear and rippling. When the water was low a bank of shingle rose in the centre. In the middle of the day the sun found its way through the leaves and struck this shingle bank, and the women liked to wade out to it and do their washing there. When the river was as low as this you could wade right across, if you knew where to step, but at other times there was a huge tree that had fallen across the river which could be used as a bridge, though it was slippery.

On the far side the ground rose sharply, and though the forest was still open and friendly at first, as soon as you reached the top of the rise it became thick and closed in around you, and the only paths were those made by animals, narrow and twisting.

Kenge showed me all this with his chest puffed up with pride. He told me I could drink water from any of the streams or from the Lelo itself; it was not like living in the village where both the food and water were dirty. He showed where the women would bathe, and warned me to keep upstream of the shingle bar, so as not to offend them. The other youths had already stripped naked and were splashing in the water, unashamed of their teen-age years, though in deference to an overclad westerner they held their hands in front of them when they turned to face me.

We went back to the camp site and Kenge started clearing the ground. Very shortly afterwards three of the great hunters arrived; Njobo the killer of elephants; Masisi the proud father; and Manyalibo the gruff, good-humoured conservative. They shouted a greeting as they came into the clearing, and looking around they threw down their hunting nets where they thought they would like to have their houses built. Njobo chose to be just to the south of the hillock, Manyalibo and Masisi, good cousins that they were, put their nets down beside each other, opposite. Then almost immediately they disappeared into the forest, each in a different direction, each with bow and arrows. In less than fifteen minutes Kenge, who had been examining the tracks of a huge buffalo that had passed by the night before, suddenly stood still and motioned me to be quiet. With his hands he mimed an antelope running, and he pointed into the forest behind Manyalibo's chosen site. Almost immediately there was a shout from Njobo, and Kenge bounded across the camp, picking up a hunting net as he went. Not more than fifty yards away we found Njobo, standing, tense, at the end of a fallen tree, peering inside the hollow trunk. Masisi and Manyalibo came running up through the glade, and taking the net they quickly put it up around the tree trunk while Njobo hurried off and returned with his spear.

Looking inside the black, gaping hole I saw two eyes, glowing a bright red and unblinking, staring at me from the darkness. Njobo had seen a *sondu*, a highly coveted kind of antelope, and had cornered it in a sharp bend of the river so that it had been forced to take refuge in the tree trunk, trapping itself. All this had taken place so silently that only Kenge's sharp ears had picked up the noises. When the net was firmly fixed Manyalibo stood on the tree trunk and thrust his long stick inside, keeping himself well clear of the opening. The forest antelopes have small but sharp horns, and when trapped they are vicious and dangerous. The *sondu*, as expected, came bolting out, snuffling with anger and pain, so fast that he tore the mesh of the net when he hit it. He lay there struggling and bleating, but only for an instant. All three hunters were on top of it, and two of them held it down while Njobo cut its throat with the blade of his spear. He looked up at me happily and said, 'This is going to be a good camp, for the forest has given us meat before we have even built our houses.'

Within another fifteen minutes the first of the women arrived. They had heard the thunder and had hurried, travelling (as they can when they want to) as fast as any man in spite of the loads on their backs and the children on their hips. They were singing as they strode into the camp with long, graceful steps, and just as the men had done they cast a quick look around and each chose a spot where she would build a house for her family. Njobo's wife, pretty young Masamba, was amongst them, and she went to where Njobo had put his net. Without stopping for a rest she threw down the load from her back, took out her knife and went off to cut the branches and leaves needed to build their new home. Njobo shouted after her that she should get enough to build my house as well, saying that we would help her by collecting more leaves. He called to Masisi and Kenge, and I followed the three of them, wading across the Lelo and into the forest beyond.

The afternoon was already half spent, and by the light we could tell that the sky was clouding over. Njobo knew where there was a large patch of *mongongo* leaves some distance off, but if we hurried we could get there and collect enough for all of us, he said. He looked at me doubtfully. But Kenge, ever practical, said that I had long legs, so surely I would be able to keep up with them, and I would be useful as I could carry twice as many leaves as all of them put together. I did keep up, but only just. Because of the time and the approaching rain they followed small antelope trails instead of the larger buffalo trails, and in this way kept to a more direct route. But it meant that even they had to stoop as they ran along, and as they did not bother to cut the branches above their heads I had to go practically on hands and knees in places. Kenge was in front of me, and every now and then he warned me by a gesture of a thorny vine, so thorny that if you ran into it unawares, it could tear your skin to shreds.

We ran without stopping for half an hour, arriving at a thick growth of *mongongo* which stretched along the hillside for half a mile or more, down to the banks of the stream below. It was like a forest within a forest, the giant stems ending in the large, heart-shaped leaves above our heads. For a whole hour we cut leaves, working quickly and silently, cutting each leaf with the knife in the right hand, leaving a stalk about four inches long, adding it to a bundle in the left hand. The others had to bend the plants down to reach the leaves, but even so they quickly accumulated small bundles and when they could hold no more they simply put them down on the ground and moved on. Yet when we had cut enough, and were at the far edge of the *mongongo* grove, they knew exactly where each bundle had been put, and retraced their steps gathering the leaves as they went. Only then, when all the leaves were collected together, did they relax and talk, and pick a few wild berries with which to refresh themselves. Then they cut lengths of vine and softened it by running it quickly from hand to hand, pulling it sharply through the fork of the thumb. They laughed when I tried it and cut myself. With this vine they tied up the bundles, gave me the largest, and set off on the way back, as fast as we had come.

As we came through the glade on the hillside above the Lelo we heard the voices of women and the laughter and shouting of children from the camp across the river; and once we got down lower and were wading through the cool stream, washing off all the dirt and perspiration, we could see long columns of smoke spiralling lazily upwards through the trees. The camp was coming to life.

The clearing was full of activity. About twenty families had arrived, and each had already put up the framework of its little house.* Arobanai and Loku, Ekianga's two senior wives, put their houses side by side at the north end of the camp, leaving a space for Ekianga and Kamaikan when they came out. Masisi's eldest son, Ageronga, helped his wife build their house between Masisi and Loku. Old Sau had already almost finished her hut between my hillock and Arobanai, but closer to Arobanai, on the far side of the path that led off to the Lelo. Masamba was busy finishing the frame for her house, mine was complete, all but a few extra sticks and leaves needed for the doorway. She had finished that first, and now she squatted down making her own home, driving the saplings into the ground with sharp thrusts each time in exactly the same place, so that they went deeper and deeper. When she had completed a circle she stood up and deftly bent the *fito* over her head,

* See plan on p. 32

twisting them together and twining smaller saplings across forming a lattice framework.

When she had finished, she took the leaves we had collected and slit the stalks towards the end, like clothes pegs, hooking two or three of them together. When she had enough she started hanging them like tiles on the framework, overlapping each other and forming a waterproof covering. There were still leaves left over when she had finished, so she let other women come and take them for their houses.

Everyone was working at something, chattering and making fun of the late-comers who had not yet finished leafing their huts, saying how wet they were going to get when the rain came. But I noticed that those who had already finished, like Masamba, even went back to the forest to cut more wood and leaves to help the lazy stragglers who had spent too much time gathering food and eating it on the way in. Sometimes there were as many as four women working on one hut, some hanging leaves from the outside, working upwards, others preferring to work from the inside, pushing the leaves through the lattice and working from the top down. Once in a while there was a loud wail as some playful child got in the way and was soundly slapped by his mother.

Just before dusk there was a sudden darkening and a wind swept through the tree-tops making leaves and small branches fall to the ground, spinning and fluttering lightly. Then with a sudden hissing sound, but loud and ominous, the sky seemed to open and the rain came pouring down. Unlike the village, where it comes down so hard that it can almost hurt you, stinging the skin and blowing into the eyes and nose and mouth, here in the forest its fall was broken by the trees and by the time it reached the ground it had spent its energy.

There were angry shouts from the men as the lightning flashed and the thunder rumbled on and on, and they peered anxiously from the tiny entrances to their huts to see when it was going to end. If it went on too long it would spoil the hunting for tomorrow. My hut was filled with the bachelors, all of whom Kenge had grandly invited in. He explained that they were too old to sleep with their parents, and not being married they had no huts of their own. We had brought our fire into the hut when the rain started, and with the heat of that and of about fifteen bodies I was nearly suffocated. Every now and then the rain found its way through the leaves above and sent a tiny waterfall cascading to the floor. Then someone would stand up and inspect the roof carefully, decide which leaf should be moved, and pushing his fingers through the lattice adjust it. If he moved the wrong leaf it started a leak in another place bringing howls of annoyance from whoever was below. Kenge said it was always like this with a new hut, that was why it was so bad not to be married. If one was married then naturally your wife would get up when it rained at night and fix the roof. But once the leaves had settled down, not even the hardest rain would come in until the leaves got old and dried out and began to curl. But tomorrow he said he would cover the roof with dead wood and branches to hold the leaves flat.

The rain only lasted for half an hour, and as soon as it was over everyone came out of their huts, building the fires up outside again, the women making a few adjustments to the roofs where they had found bad leaks, the men wandering off idly with their bows and arrows to see if they could find a bird or monkey for the evening meal before it was too dark. The huts steamed as they dried out, and a blue haze of smoke hung above the camp and was coloured suddenly with orange and gold and red as the clouds parted and the sun on the invisible horizon sent a last blaze of light across the sky. The birds burst into a final chorus of song, and two women argued loudly with each other as to who should have the rest of the leaves that Masamba had cut to help them.

Everyone was tired from the long day's trek and from the flurry of building camp so quickly, and as soon as it was dark and the evening meal had been eaten we all crept into our huts and went to sleep. Kenge, seeing that I was a little concerned, had turned some of the bachelors out, so there were only nine of us. For an hour the others lay around the fire, while I was pushed into a corner, and discussed where they would go to hunt the next day and with which girls they would flirt. They poked sleepily at the fire, sending up showers of sparks and filling the hut with smoke. In all the huts around us families

65

were chatting amongst themselves in a drowsy way, and occasionally some eavesdropper shouted across a suitable comment, bringing a burst of laughter from all the nearby huts. The leaf walls keep out no sound at all, and anything said in an ordinary talking voice can be heard four or five huts away. But before long the voices got sleepier and sleepier, and finally the only sounds were the chirping of crickets, the soft swishing of the already swollen Lelo, and the quiet falsetto of Makubasi as he sat outside his hut, all alone by a blazing fire. He held his baby son on his lap, and he sang a lullaby, half to himself and half to the child. He was still singing when I went to sleep.

The next morning the camp was up at dawn, and before it was properly light there was hardly a soul left about—only old Tungana sitting where his son had sat the night before and, like his son, singing to the baby on his knee. Kenge said that some of the men had gone hunting early as they needed a lot of meat to begin the *molimo* festival for Balekimito and they did not want to start until they had plenty of food. The rest of the men and women were out cutting the leaves and saplings needed to finish building the camp. We went off to join them, not far off, and we spent the rest of the day making the simple furniture of a pygmy household.

Sticks lashed together and bound by vine thongs to a frame make one of the most uncomfortable beds in the world. The sticks themselves are comfortable enough, but for some reason these beds are made so that each end bends sharply upwards and however you lie on it the effect is worse than being in a badly slung hammock. Some pygmies don't bother with a bed but just put the sticks on the ground and lie on them, or on dry leaves. Tables are made the same way, the frame resting on four forked sticks, but we were the only household to bother with a table. The others would build them later, Kenge said, when the hunting brought in so much meat that they could spare some to dry for the villagers. Then they would build tables, higher than ours, and light fires under them.

Chairs were even simpler to make. Kenge cut four sticks, each about three and a half feet long, and twisted a vine thong around the middle of the bundle. He stood them on the ground, and with a turn of his hand he splayed out the ends and had a comfortable seat. When you get used to this kind of chair you can even go to sleep, resting your back along one stick and hooking your feet over the opposite one. Pygmies, in the forest, are clean people, and they do not like sitting on the ground. They either make these simple chairs or they sit on logs, often on the end of a log sticking out of a fire. If the worst comes to the worst they will pull a leaf from the roof of the nearest hut and sit daintily on that.

During the day more families arrived, and while they were busy building their huts, those who had arrived the day before put the finishing touches to theirs. Sometimes someone would apparently have a grudge against a neighbour, or a dislike for someone who had built their hut opposite, and I saw several women adding to the entrances of their huts, pointing them in different directions, towards their friends or relatives. During the course of any one camp women are continually adding to their huts or changing the way they face, and I found that a good way of keeping track of the little jealousies that exist in any small community was to make a daily plan of the camp, noting which huts were being altered to face in which direction. I noticed that Arobanai, Ekianga's senior wife, faced her hut very deliberately away from old Sau, and that nobody was anxious to make use of the space between Sau's hut and mine to build there.

On the third day Ekianga arrived with Kamaikan, and they installed themselves with the rest of the family at the north end of the camp. Shortly afterwards Cephu arrived. Ekianga had brought news that he was on his way, and there had been a lot of argument. In spite of the auspicious killing of a *sondu* on the first day, the hunting since had not been too good, and already the camp was large. There were about twenty-five huts now, forming two circles, each side of the central barrier of brush and trees. In the early afternoon there was a silence in the camp and Kenge stood listening, looking across the camp to the south. He turned and told me that Cephu was not far away and would be here in a quarter of an hour. Almost to the minute Cephu arrived, leading a band of another five families. Most of them stopped at the edge of the camp at the south end, only Cephu and one other came over to the main camp to greet us and announce his arrival. A new fire had been built in the centre of our clearing, opposite Manyalibo's hut, and Cephu sat down there, one of the younger men getting up to give him his chair. He asked about the hunting and gave us news of the village. He said that his brother, who was a cripple, had not wanted to come so far and had made a small camp half-way, near Apa Kadi-ketu. After a few minutes he left, not being offered anything to eat or any tobacco to smoke, and started building his own camp about fifty yards to the south of the extreme edge of the clearing. He built a tight little circle of huts connected by a path to the main trail on one side, and to the camp on the other. He chose that site because at that end of our camp he had some relatives.

The same day Manyalibo's daughter also arrived, with her husband, Ausu. Ausu was a handsome pygmy, strong and clever at hunting both with spear and with net. He had caught many wild pig, although he was only in his early twenties, and once, when still in his teens, he had nearly been killed by one. His wife, Kondabate, was a pygmy belle if there ever was one, and she knew it. The two had been married for nearly three years, but had no children. Kondabate still had her fine maiden's figure, with firm, upstanding breasts, and nobody ever mentioned the fact that they were childless. Kondabate decorated herself more than any other pygmy girl I have ever known, and by every possible means, negro and pygmy. She not only dyed her body with designs of vines and fruits, painted on with the black juice of the gardenia fruit, but she also had elaborate patterns cicatrized on her stomach, and her teeth were filed to sharp points so that when she smiled or laughed, as she frequently did, she looked just a little dangerous. Kondabate decided to build her house to the north of the hillock, so that she was next to old Sau, and opposite her father.

During the afternoon Amabosu, Sau's son—the great singer —went around from hut to hut with Manyalibo's adopted son, Madyadya. Manyalibo had offered his *molimo* for Balekimito's festival, as Ekianga said he had none. The two youths stood outside the entrance to each hut with a length of cord made from *nkusa* vine, the kind of cord used for making hunting nets. Madyadya held one end while Amabosu tied the other in a noose and laid it in the doorway of the hut. The owner of the hut, or her husband if she was not there, put a small bunch of bananas in the noose, or a manioc root, or a bundle of mushrooms, or perhaps even a small piece of meat, and Madyadya would jerk the string taut. After a brief mock struggle he would haul in the catch and put it in his basket, then go on to the next hut. From every hut the two collected an offering of food in this way, and when their basket was full they hung it up on a tree stump by the central fire that was now blazing between Manyalibo's hut and the hut his daughter was busily making. This fire had itself been made up from offerings of wood collected from each family hearth, I remembered seeing Manyalibo go around taking an ember or log from every fireplace whether the owners were there or not. I noticed that only men sat at this fire, and I asked Kenge why this was.

He told me it was the *kumamolimo*, 'the place of the *molimo*', and now that everyone was here we were going to start the festival. The food that had been collected from hut to hut was to feed the *molimo*, because the *molimo* was a hungry animal that ate a great deal. He added, pointedly, that it smoked a great deal too, and when I gave him several packets of cigarettes he solemnly put them with the food in the *molimo* basket. Then I went back to my hut.

About an hour later I noticed that Makubasi and Ausu were standing outside, obviously wanting to say something. Usually they were not backward about interrupting, and I was puzzled when Ausu beckoned with his head that I should come outside. He whispered in my ear that a few of them were going off to get the *molimo*, as we were to start the festival that night. Masimongo, Kenge's brother-in-law, had said that he wanted to bring out his *molimo* as well as Manyalibo, and make it a doubly great festival. Half a dozen youths were waiting to see if I would go with them, but the place where the *molimo* lived was a long way off and we would have to run all the way, they told me, and return when it was dark. Of course I said I would go, and we started off immediately. But half-way across the camp Masimongo leapt in my way and shouted that I was not to go. For a moment there was a loud and sharp argument, most of which I could not follow, but finally Njobo, to whom everyone

69

always listened, said that he had seen me in the forest and knew I could travel as well and as fast as the others, so that I would be safe. Masisi objected with a shrill torrent of words, saying that the *molimo* was none of my business, and for a moment I felt like backing out and saying that I would not go, as I certainly did not want to cause any bad feeling.

The others were already moving off at a run, down the path through the glade, not waiting for me. Then Njobo pushed me on the shoulder and said to follow them quickly, if I was sure I could keep up and would not be afraid of travelling in the forest in the dark. He said the others would not be able to wait for me if I dropped behind, this was a serious matter. As I hesitated he added, in a loud voice directed at everyone in general, but particularly to Masimongo and Masisi, that the *molimo* had already sung to me before, and no harm had come of it. I was after all, an adult, and of the forest, so there was no reason why I should not go with the others. He was still talking as I caught up with Makubasi, who was hanging back to see if I was coming. We jumped into the Lelo, more swollen than ever, and struggled across. The water came up to my waist, nearly up to his shoulders. It filled my shoes with sand and grit, but there was no time to take them off. As we made our way through the water the others ducked their heads and washed themselves, and I did the same.

When we reached the far side, Ausu, leading, was half-way up the hill while I was still splashing around trying to find a foothold in the slippery bank. The forest was not yet dark, but it was very silent, and for once none of us made a sound as we ran along, following the antelope trails, towards the north.

Chapter Four

THE TRAILS we followed were narrow, and in places even the pygmies had to stoop and bend. We ran at a steady pace without slowing up once, and never hesitating as to direction even though there were times when it was difficult to see any trail at all. When we climbed upwards we went faster as it was easier to keep a footing that way. I was the only one whose feet made any noise, the others ran so lightly that they barely touched the ground, but rather seemed to skim along just above it, like sylvan sprites.

After about forty-five minutes we came to a stream that was larger than the others we had crossed, though it was only a few yards wide and very shallow. For the first time we slowed to a walk and stepped through the bubbling water, lifting our feet high so as not to splash. The forest was silent now, except for the evening chorus of crickets and frogs, and on the far side of the stream we halted while Ausu and Makubasi looked around and listened, to reassure themselves that we were alone.

I had no idea of how far we had come or in what direction, but I knew we had left the camp far behind and it was the first time I had known a group of pygmies to be so silent. Normally, unless hunting, they are deliberately noisy; but now, just at the time that leopards would be prowling about in search of food, they seemed unwilling to disturb the forest or the animals it concealed. Just the opposite, in fact. It was as if they were a part of the silence and the darkness of the forest itself, and were only fearful lest any sound might betray their presence to some person or thing not of the forest.

They stood there, quiet and still, and it struck me with a sudden shock that not one of them carried a spear or bow and arrow. As they peered into the dusk and cocked their heads first on this side then on that, satisfying themselves that we were really alone, it seemed that they felt themselves so much a part of the forest and of all the living things in it that they had no need to fear anything except that which was not of the forest. One of them said to me, later, 'When we are the children of the forest, what need have we to be afraid of it? We are only afraid of that which is outside the forest.'

We stood there, water from the stream still running down our legs, until we were satisfied that all was well. Then Makubasi nodded to Madyadya, Manyalibo's adopted son, and to another boy. They ran on ahead, and two others followed them after a whispered discussion. The rest of us walked slowly on until we came to a small clearing nearby, the far side of which was only just visible as the last glimmer of daylight dwindled away into blackness. We stood there waiting, saying nothing. Both Ausu and Makubasi wandered about aimlessly, poking here and there from habit to see if there was any fruit to be had for the picking, but it was too dark to see.

Just as I was about to ask where the others had gone they reappeared, announcing their presence with low whistles that sounded like the call of the night bird. They were in two pairs, each carrying between them, over their shoulders, a long, slender object. Even at that moment I wondered if they would veer off into the complete blackness of the forest before I could see more closely, but they came on towards us, Madyadya carrying the rear end of what proved to be a huge tube of some kind, fifteen feet long. He gestured proudly and said: 'See, this is our *molimo*!' Then he turned and, putting his mouth to the end of the trumpet, as it was, he blew a long, raucous raspberry. Everyone doubled up with laughter, the first sound they had made since leaving camp.

I was slightly put out by this sacrilege, and was about to blame it rather pompously on irreligious youth, when I saw something that made me even more upset. I do not know exactly what I had expected, but I knew about *molimo* trumpets, and knew that they were sometimes made out of bamboo. I suppose I had expected something elaborately carved, decorated with patterns full of ritual significance and symbolism, something sacred, to be revered, the very sight or touch of which might be thought of as dangerous. I felt that I had a right, in the heart of the tropical rain forest, to expect something wonderful and exotic. But now I saw that the instrument which produced such a surprisingly rude sound, shattering the still-

72

ness as it shattered my illusions, was not made of bamboo or wood, and it certainly was not carved or decorated in any way. It was a length of metal piping, neatly threaded at each end, though somewhat bent in the middle. The second trumpet was just the same, shining and sanitary, but only half the length. Everyone was looking at me to see what my reactions would be. I asked, keeping my voice low, how it was that for the *molimo*, which was so sacred to them, they should use water piping stolen from roadside construction gangs, instead of using the traditional materials.

But evidently now that they had got the trumpets in their possession there was no longer any need for silence, and they answered calmly and loudly with a counter-question, 'What does it matter what the *molimo* is made of? This one makes a great sound, and besides, it does not rot like wood. It is much trouble to make a wooden one, and then it rots away and you have to make another.'

Ausu, to prove how well it sounded, took the end of the longer pipe, which evidently belonged to Manyalibo, and all of a sudden the forest was filled with the sound of trumpeting elephant. The others clapped their hands with pleasure and said, 'You see? Doesn't it sound well?'

My conservative feelings were still wounded, however, and it gave me some pleasure to see the difficulty the pygmies had on the way back, carrying a fifteen-foot length of drainpipe through the forest. It was pitch dark now, though the moon was full, yet we still followed only the smallest animal trails, this being the quickest route. Frequently Madyadya, now carrying the front end of his father's trumpet, turned too sharply, following the trail, so that whoever was carrying the rear end was forced off the path and the trumpet became jammed between two trees. In spite of this we returned almost as quickly as we had come.

When we came to the first stream we all stopped. The moonlight was just strong enough for me to see both trumpets laid carefully in the water, then lifted up so that the water ran through from end to end. 'The *molimo* likes to drink,' I was told. We washed our feet and continued, but at every stream we came to we went through the same performance, both trumpets being immersed and allowed to 'drink'. Every now and again, whoever was at the rear end of either trumpet managed to find enough breath to blow a long blast that was sometimes musical, but more often not. The younger pygmies all seemed anxious to have a turn but Ausu, who was the only one who really knew how to play, kept aloof. He ran ahead of us all, silently, into the night.

It was so dark along the trails that I had to run barely two feet behind whoever was in front, occasionally reaching out and touching him to make sure he was still there. I had learned to lift my feet high to avoid roots, but I did not know the trails inch by inch, as did the pygmies, and once I put my foot in an animal hole and fell. But nobody stopped, and I had to run all the faster to catch up. Only at the streams was there just enough light for us to see each other dimly; and at each stream I looked at those two drainpipes being lovingly washed and fed with water. I was curious as to how far this was a ritual act, and how far it was originally a practical measure that would make a wooden trumpet swell, closing any possible cracks, so that it would sound better, Ausu said that of course that was why they did it, they did the same thing to all their wooden whistles and flutes as well; it made the sounds stronger. Then he added, casually, 'We know it makes no difference to the sound of the metal trumpet, but we do it anyway.'

When we reached the top of the hill that runs gently at first, then steeply down into the Lelo River, we all halted and stood in the glade. It was lit with silvery patches of light, shining here and there uncertainly where the moon found its way down through the rustling trees. Even the darker shadows glowed every now and then, but with a strange, natural luminosity that came from the mushroom-shaped nests the termites build up against tree trunks. We could hear the Lelo swishing by below, and on the far side of the river voices from the camp floated lazily across to us, as though we were gods and they were mortals, infinitely far away.

We stood there for a moment, then Ausu took the end of the long trumpet, with Madyadya holding the front end on his shoulder, pointing it over the river towards the camp. Once again the night was shattered as though there was a whole herd

of elephant trumpeting shrilly all around us. When the echoes
died away, the camp across the river had fallen silent. Ausu
blew into the trumpet again, and this time it was as if a leopard
were growling angrily nearby, and one of the youths clapped
himself under the armpit, as pygmies frequently do, and said
to himself: 'That is its voice, just exactly.' He clapped himself
again with delight, and gave a little growl of his own.

Little more than a minute passed when Amabosu, the great
singer, appeared from the shadows below us. He looked at us
without saying anything, without smiling or making any greeting. His eyes had the strange look they often took when he was
singing or playing the drums in a dance in the village. They
seemed to be seeing something quite different from what was
going on in front of them. He went up to Ausu and took the
molimo from him, and he gently filled the forest with strange
sounds; the rumblings and growls of buffaloes and leopards, the
mighty call of the elephant, the plaintive cooing of the dove,
and interspersed were snatches of song, transformed by the
trumpet into a sound quite unlike the song of men—richer,
softer, more distant and unapproachable.

He continued blowing as we moved off, down the slope and
to the bank of the river, but when we plunged into the water
and waded across the two trumpets left us and were carried upstream. They growled and bellowed somewhere behind us as
we made our way along the path to the camp. In the clearing
the men were standing up to see us arrive, and the women were
hurrying about doing what they had to do before shutting
themselves up in their huts. The voice of the *molimo* had warned
them that it was time to retire. The children had all been chased
inside, and one by one the leaf doors were pulled across the entrances to the little huts, and only the men were left outside.

Njobo was there to greet us, and the first thing he asked was
whether I had been able to keep up with the others, or if I had
'walked like the BaNgwana'. He then turned to the others with
an obvious, 'I told you so' air and said, 'You see, it is all right,
he knows how to walk.' This apparently slight compliment was,
in fact, a very real one, because to the pygmy it is one of the
most important differences between the people of the forest and
the people of the village that the latter do *not* know 'how to

walk'. Walking, to the pygmy, means being able to run swiftly and silently, without slipping, tripping or falling. Every day he depends for his food on his ability to 'walk', and more than once his life will be saved by the same ability, when he has to run from a charging buffalo or creep away unnoticed from a sleeping leopard.

Masisi, unbelieving, asked if I had not fallen at least once and when he was told I had he roared with laughter. Masimongo, who had been another to object to my going, pointed to my torn clothes, dripping with water and perspiration, and filthy where I had fallen, and he too laughed. Nobody mentioned the *molimo*.

Kenge came up and asked disapprovingly why I had gone off so quickly without telling him, saying he had cooked dinner and now it was all cold and I would have to go without. He gave my clothes a withering look and said he would wash them out the next day, and that I had better take them off now and get dry. I went over to my hut and changed, and when I came out he had a piping hot meal waiting by a blazing fire.

What followed that evening was to follow every evening for the next two months. It was the focal point of the *molimo* festival, and the repetition every night was an essential part of it. The ceremony, if you can call it that, did not centre around ritual objects, as do the ceremonies of the local negro tribes, nor did it follow any prescribed ritualistic detail beyond a general overall pattern. It was plain that the *molimo* trumpets were not considered as highly sacred objects in themselves; what was important, for some reason, was the sound they produced. It was also very clear that the people of the forest wanted to be quite sure that the village world was well sealed off, because before the men gathered together that evening I found that the path leading from the camp towards the village had been closed. A fallen branch and some logs had been dragged across it, shutting out the profane outside world.

The *molimo* of the pygmies is not concerned with ritual or magic. In fact it is so devoid of ritual, expressed either in action or words, that it is difficult to see what it *is* concerned with. Every day, around midday, a couple of youths would go around the camp, just as they had done this day, collecting offerings of food and firewood from hut to hut, for the *molimo* concerns everyone, and everyone must contribute. And each evening the women and children shut themselves up in their huts after the evening meal for the *molimo* is the concern mainly of the men. And when the women have retired the men gather around the '*kumamolimo*', 'the place of the *molimo*', and sit gazing into the *molimo* fire. Nearby a basket hangs, full of the offerings of food that will be eaten later. But first the men must sing, for this is the real work of the *molimo*, as they say; to eat and to sing, to eat and to sing. Yet behind these apparently simple outer trappings of the festival there was an atmosphere of almost overwhelming expectancy.

That first night, after I had eaten my meal, I went over to where the men were already gathering at the *molimo* fire, halfway between Manyalibo's hut and that of his daughter, Kondabate, and her husband, Ausu. I noticed that Amabosu, the singer, was not there, and I knew why when a few minutes after the singing had begun I heard the voice of the *molimo* answering, way off by itself in the forest. It no longer worried me that the trumpet was a metal drainpipe instead of a piece of bamboo or wood, because now that I could not see it I realized that Ausu was right. It was the sound that mattered. As the men sang in the camp, the voice of the *molimo* echoed their song, moving about continually so that it seemed to be everywhere at once. During a lull in the singing it started giving animal growls, and the men looked around to make sure that all the women were safely in their huts. Almost immediately Amabosu and Madyadya came running into the camp from the far end, past Cephu's camp, picking their feet up high in the air so that more than ever they seemed to be floating above the ground they trod. They came right up to the fire and squatted down either side of it, the front of the trumpet still held on Madyadya's shoulder as Amabosu warbled softly into the other end.

I noticed that the trumpet was still wet, it must have been given a 'drink' just before it was carried into the camp. Now both youths stood up, and with Amabosu singing all the time, in reply to the song of the men, Madyadya took the end of the trumpet in his hands and waved it up and down, passing it through the flames of the *molimo* fire and over the heads of all of us sitting there. Some of the men took up hot ashes and rubbed them over the trumpet, and some even put live coals in at the far end. As Amabosu sang harder and louder sparks flew out over Madyadya's shoulder, disappearing into the night with the song.

Then the two of them went dancing off, around the camp once and into the forest. There were a few last defiant bellows from the direction of the river, and no more was heard. The men stopped singing and prepared to eat. Two cauldrons had been boiling over the fire, and both were now ready—one contained rice, the other a kind of spinach made from the leaves of the manioc plant. Eaten together they are delicious. I wondered if there would be a precedence as to who should eat first, or if there would be any ritual, but there was none. Everyone ate his fill as and when he pleased, a couple of the older men saying they would wait until the youngsters had finished before they ate. But they did not wait for long. I was told, however, that everyone *had* to eat, just as no adult male was allowed to sleep but *had* to sing while *molimo* singing was in progress. Apparently one of the greatest crimes that a pygmy male can commit, if not the greatest, is to be found asleep when the *molimo* is singing. One of the men stood up, and miming as all pygmies love to do, he showed how they used to search for sleeping men, a spear under each arm, and how if they found one they would spear him in the stomach and kill him completely and for ever. I was told that then his body would be buried under the *kumamolimo* and nobody would be allowed to mention his death, or even remark on his absence. The women would be told that the *molimo* itself, the great animal of the forest, had carried off one of their number, but like the men they would ask no questions and would never again mention the missing man.

The *molimo* was often referred to as 'the animal of the forest', and the women were meant to believe that it really was an animal, and that to see it would bring death. That, of course, is why they were all bundled off to bed with the children before the trumpet was ever brought into camp. And even when it *was* brought in it was often shielded by a number of youths so that if any woman should happen to look, she would see nothing. The animal sounds it produced were certainly realistic, but I wondered what the women thought when it sang. What kind of animal was it that one moment could make such threatening sounds, and the next instant sing more beautifully than anything else in the whole forest?

I remembered what Ausu had said about the only important thing about the trumpet being the fact that it had a good voice and could sing well, and I was reminded of the pygmy legend of the Bird with the Most Beautiful Song. This bird was found by a young boy who heard such a Beautiful Song that he had to go and see who was singing. When he found the Bird he brought it back to the camp to feed it. His father was annoyed at having to give food to the Bird, but the son pleaded and the Bird was fed.

The next day the Bird sang again, it sang the Most Beautiful Song in the Forest, and again the boy went to it and brought it back to feed it. This time the father was even more angered, but once again he gave in and fed the Bird. The third day (most pygmy stories repeat themselves at least three times), the same thing happened. But this time the father took the Bird from his son and told his son to go away. When his son had left the father killed the Bird, the Bird with the Most Beautiful Song in the Forest, and with the Bird he killed the Song, and with the Song he killed himself and dropped dead, completely dead, dead for ever.

There are other legends about song too, all telling how important it is, and I wondered just why. I looked at the others sitting with me around the *kumamolimo*. The flames from the fire flickered and lit up their serious faces, shining brightly in their large, honest eyes. Some gazed straight into the fire, others leant back and looked up into the trees as they sang. A few lay down on their hunting nets, but they were awake and they sang with the rest. Then I saw that Makubasi had his infant son on his lap, holding the boy tightly against him, wrapped in his powerful arms, rocking backwards and forwards. He was singing quietly, with his mouth up against his son's ear, and the words of the song, like the words of most *molimo* songs, were few. They simply said, 'The Forest is Good.'

The pygmies seemed bound by few set rules. There was a general pattern of behaviour to which everyone more or less conformed, but with great latitude given and taken. Even in matters of this sort, where all women and children were meant to be abed, if Makubasi felt like getting his son and singing to him at the *kumamolimo*, nobody objected or thought anything of it. Possibly this was because the baby was too young to know what was going on, and could not tell the others what he saw, possibly simply because it did not matter very much anyway, for the Song was not only Beautiful, it was Good and Powerful.

It took me some time to realize just how important song was. That first night I just thought of it as being a beautiful sound. After we had eaten the food that had been offered to the *molimo* we sang again, and at one point the younger men got up and danced around the fire, excitedly, but without any of the frivolity that would come later. This was the first night of the *molimo* and everyone was on his best behaviour. Then when the dancing was over we heard the trumpet again, and as we sang it sang back to us, getting further and further away. As it went it carried our song with it, deeper and deeper into the forest. Finally we could hear it no more, only the echoes of our own voices came back to us from the night.

Not long after that old Moke stretched and yawned and said that he was tired and that everyone should go to sleep and let him sleep. But the younger men paid no attention and went on singing, so Moke resignedly joined in with them. But after another half-hour, when it was not very long before dawn, even the younger men decided to quit. Moke picked up the log he had been sitting on, the end of which was glowing in the hot ashes, and shuffled back to his hut, swinging the log so that it glowed brightly and lit his way across the camp. Manyalibo simply stretched out where he was, right beside the *molimo* fire, and went to sleep there. A number of youths did likewise, their hunting nets making comfortable beds, but I went back to my hut. I found Kenge there, fast asleep. He was on my bed, wrapped in my blanket, snoring contentedly, and the hut was filled with smoke from the fire he had lit beside him.

*　　　　*　　　　*

In spite of the hunting that had to be done each day, the men sang in this way every night. And every night we looked to see who would come from Cephu's camp, for the amount of interest that Cephu showed in the *molimo* was a topic for constant discussion and criticism. Masimongo, who had relatives in Cephu's camp, always upheld him when he did not appear himself or when his little group made a poor showing. But Manyalibo allowed no such excuses, and his voice carried much more weight. He said, good conservative that he was, that all men should take part in the *molimo* regardless of what camp they belonged to, if they were anywhere near at all. Then he went through the mime showing what would have happened in the old days to anyone who did not attend, and who was found sleeping. I thought momentarily of Kenge.

But it was not only a matter of singing late into the night. Each morning, too, there was activity, though of a different nature. The evenings were serious, the older men taking the lead in the singing and leaving the dancing largely to the younger men. Sometimes there was a lot of horseplay, but on the whole the evening singing was obviously a serious matter and the older men like Moke kept an eye on what was going on and held the youngsters in check. Often the *molimo* fire was made into two fires, the older men sitting at one and the youngsters at the other, and then all the dancing went on at the youths' fire while the elders sat at theirs and just sang, and watched their sons with dreamy eyes.

But in the mornings the youths came into their own completely. Before the first glimmer of light filtered through into the glade those of us who had managed to get to sleep were woken up by a violent and raucous trumpeting from just outside the camp. This brought cries of protest from the older men, some of whom appeared sleepily at the entrance to their huts and shouted, 'Animal! Go away and keep quiet and let me sleep!' but to no avail. From the far end of the camp, near the path leading to Cephu's clearing, a wild cavalcade of youth swept into view, shouting and yelling, clustered so thickly that it was impossible to see the trumpet they were carrying. But it was there, with Amabosu blowing it. It blasted and shrilled and growled and bellowed, and it rampaged around the camp, overturning any of the crude chairs that had been left outside, scattering the remains of fires in all directions, and beating on the roofs of huts to wake everyone up. Sometimes the youths seemed to divide at random into two sides and play a sort of tug-of-war with the trumpet, some trying to lead it in one direction, some in the other. It was not a real contest of strength, only acting, but I wondered if perhaps this represented the original struggle for possession of the *molimo*, a struggle that some say was between pygmies and animals, or between men and women as the story is otherwise told.

Ekianga took a prominent part in these morning affrays. He used to lead the youths, always holding two spears, one under each arm, pointing down at the ground as though he were looking for any sleeping person who should have been taking part in the activity. Sometimes he acted it out, stopping as though he saw an offender. Then he danced about and finally jabbed the spear blades at the earth. I said to Kenge that this was what would happen to him if he went off and took my bed and blanket again before the evening singing had finished. Kenge laughed loudly and shouted across the camp what I had said. But then he was serious and said that he had been singing all the time until Moke had first said he wanted to go to sleep. 'When the old men want to sleep it is all right for us to sleep,' he said. Then he added, laconically, 'And of course if we are tired it is all right for us to go to sleep, naturally.' He thought the conversation not very interesting, so he gave an animal call and ran with great leaps and bounds, like the antelope he was named for, across the camp to join in the tug-of-war.

Sometimes, if anyone had offended the previous day, usually by being too argumentative, the youths in the morning rampage paid particular attention to the offender's hut. Cephu's camp, of course, came in for the most attention, and although Cephu complained loudly each time, at the time, he never brought it up later as an issue for discussion or dispute. Once, the morning after two brothers had been fighting, the 'animal of the forest', making more noise than ever circled all around the hut of one brother and finally pounced on it, beating on the roof and tearing off leaves and sticks. Some of the youths climbed up a tree hanging over the hut and broke off a heavy branch so that it fell on the roof, blocking the entrance, but doing no real damage. Inside, Masalito and his wife screamed their protests, and then there was a sudden silence. Masalito had done the most dreadful thing of all, he had told the youths to take 'that animal' away and throw it back in the water and stop all their noise. This spoiled the whole illusion, which is only a pretence in itself, that the women think that the *molimo* is an animal and do not know that it is a trumpet surrounded by a lot of noisy youths. The silence was followed by cries from all over the camp; cries of shame shouted by both men and women. The 'animal' was galvanized into even greater action and Masalito's hut was in danger of being completely destroyed when Njobo came sleepily out into the clearing and told the youths to go away and leave Masalito alone. They had probably had their fill by then anyway, and they left, making a few last defiant noises, some in the direction of Njobo. As soon as they were gone the women came out of their huts, looked around to see how much damage had been done, then went to wash themselves and get breakfast. Children scampered around an irate Masalito as he cleared up the wreckage outside his hut, and as they danced up and down I heard one of them, bolder than the others, give a tiny hoot in imitation of the trumpet. Masalito tried to grab the child, but they all ran away, laughing, to torment someone else.

For a month I sat every evening at the *kumamolimo*; listening, watching, and feeling; above all, feeling. If I still had little idea of what was going on, I felt that air of importance and expectancy. Every evening, when the women shut themselves up pretending that they were afraid to see 'the animal of the forest', every evening when the men gathered around the fire, pretending they thought that the women thought that the drainpipes were animals; every evening when the trumpet drainpipes imitated leopards and elephants and buffaloes; every evening, when all this make-believe was going on, I felt that something very real and very great was going on beneath it all, something that everyone else took for granted, and about which only I was ignorant.

It was as though the songs which lured the 'animal' to the fireside also invoked some other kind of presence. As the evenings wore on towards mornings, the songs got more and more serious and the atmosphere, not exactly tense, but charged with an emotion powerful enough to send the dancers swirling through the *molimo* fire as though its flames and red-hot coals held no heat, as though the glowing embers were cold ashes. Yet there was nothing fanatic or frenzied about their action in dancing through the fire. And the1 there was always that point when the 'animal' left the camp and returned to the forest taking the presence with it. I could feel it departing as the mellow, wistful voice of the *molimo* got further and further away.

The morning rampages carried no such presence with them; they seemed more tangible both in actuality and in purpose. I learned that after these rampages the trumpets were brought back to the Lelo, just above where the men bathed, and they were immersed there during the day time so that the women would not see them. One day there was a terrible storm and the Lelo rose four or five feet, transformed into a torrent. When the rain abated in the late afternoon Madyadya went to look for the trumpets and found that one of them had been washed away. This caused tremendous hilarity, though, of course, everyone tried to disguise what they were talking about in front of the women. The women, playing the same game, pretended not to know what was going on. Eventually the trumpet was found some distance downstream, full of sand and gravel and mud. But that night, in spite of its prosaic experience, it was a mysterious thing again, alive and aweful, inspiring, and wonderfully beautiful; something that made the eyes of young and old alike light up with pleasure as they heard it sing. And, as always, there was that look in Amabosu's eyes as he worked his magic with the drainpipe. He was no longer Amabosu; he had some other personality totally different, and distant.

* * *

One day, when I was beginning to wonder if I would ever find out what it was all about, I was out with Kenge and his half-brother, Maipe. We had left at dawn as we wanted to go to a distant part of the forest, about five hours' walk away if one carried no load, and return by nightfall. About half-way there, way to the north of the Lelo and near the great falls of the Kare

River, we stopped for a rest. We were in an *idu*, a small natural clearing used by the forest animals as a salt-lick. For some reason we started talking about water and I asked again why the *molimo* trumpets always drank when they crossed a stream. Maipe, who seldom said much, said again that they always gave wooden trumpets water to drink because it made them sound better, just as they do to their wooden flutes and whistles which often will not sound at all unless they have been soaked. So, he said, it had become a custom and they still followed it even if the trumpet was metal. I then asked why the trumpets were kept in water, and was again given a practical answer—that this was the easiest place to hide them. I finally arrived at the question I had been leading up to, what happened to the *molimo* trumpets at the end of a festival? Did they go back to live in the water?

This struck both Maipe and Kenge as terribly funny and they said certainly not, if the trumpets were kept in the water always they would rot right away and everyone would be spending all their time making new ones instead of hunting. Without any further prompting they went on to tell me how the trumpets are normally made, from the young, straight *molimo* tree, a tree that has a soft centre that can be bored out laboriously, using different vines and woods as drills, twisting them between the palms of the hands. This took a very long time, they said, but this was the real *molimo*, and its voice was the best of all. Njobo used to have one such trumpet, but it had rotted away and now all he had was a short bamboo one. Kenge told me that the trumpet which Masimongo owned had originally belonged to Adenota, Kenge's father, and that by rights was now his. But he said he couldn't be bothered with it. 'It costs too much, and it makes too much work. Always going around collecting food to feed everyone . . .' he shook his head in disgust. He had given it to Masimongo when Masimongo married his sister. He could have it back if he wanted, or he could make another, but it was all really too much trouble.

Maipe said nothing, but after a while, when we were moving off, he asked me if I would like to see the sleeping place of a trumpet. He led me along an antelope trail, so low that I had

85

to go almost on hands and knees for it followed the banks of a stream and the *mongongo* plants grew thickly. After a while the stream descended a slope in small cascades, and the undergrowth cleared. We stood up and saw that water was all around us, rippling over rocks and joining together at the bottom in a sizeable pool. Maipe looked around, then walked surely into a patch of undergrowth and called me. I went after him and found him at the foot of a tall tree the base of which was hidden by thick bush. But through the bush ran a vine, looped around the tree about three feet off the ground. Maipe said that any pygmy seeing that would know that it marked the sleeping place of a *molimo* trumpet and would keep away, for it did not concern him and the *molimo* should not be disturbed while sleeping. I asked where the trumpet was, and Maipe pointed up above his head. I could see nothing except the immense branches of the tree spreading out high above, more than a hundred feet up. 'That is where it sleeps,' he said. 'It is safe there. It sleeps there until it is needed.'

I asked when it was needed, but Maipe had said all he was going to say. He looked up at the tree for a last time, then motioned with his head that we should be on our way. Kenge whispered to me that when we got back he would ask Moke to tell me more. It was better for old men to talk about these things.

That same night, just before the singing started, Moke was sitting outside his hut, whittling away at a new bow he was making, talking to himself because there was nobody else for him to talk to. Kenge had spoken to him when we got back to the camp at dusk, but I did not expect anything to come of it, and Kenge had said nothing to me. Moke looked up and stared in my direction, then he looked back at his bow and continued whittling away with a rough stone. But he spoke a little louder, so that I could hear. 'Ebamunyama,' he said, using the name I had been given, 'Pika'i to . . .' 'Come here . . .' He spoke as if he was addressing the bow, and when I went to him he still did not look up, but just told me to sit down. Talking to the bow in his soft old voice he said that he had been told I was asking about the *molimo*, and that this was a good thing and he would tell me whatever he could. He went on to tell how the pygmies call out their *molimo* whenever things seem to be going wrong. 'It may be that the hunting is bad,' he said, 'or that someone is ill, or, as now, that someone has died. These are not good things, and we like things to be good. So we call out the *molimo* and it makes them good, as they should be.'

Kind, quiet old Moke, all alone and without a wife to look after him, still working away at his bow, occasionally looking along the shank to make sure he was keeping the line true; he told me many things that evening. But more than anything he told me, or rather he showed me, how the pygmies believe in the goodness of the forest. 'The forest is a father and mother to us,' he said; 'and like a father or mother it gives us everything we need—food, clothing, shelter, warmth . . . and affection. Normally everything goes well, because the forest is good to its children, but when things go wrong there must be a reason.'

I wondered what he would say now, because I knew that the village people, in times of crisis, believe that either they have been cursed by some evil spirit or by a witch or sorcerer. But not the pygmies; their logic is simpler and their faith stronger, because their world is kinder. Moke showed me this when he said, 'Normally everything goes well in our world. But at night when we are sleeping sometimes things go wrong, because we are not awake to stop them from going wrong. Army ants invade the camp, leopards may come in and steal a hunting dog or even a child. If we were awake these things would not happen. So when something big goes wrong, like illness or bad hunting or death, it must be because the forest is sleeping and not looking after its children. So what do we do? We wake it up. We wake it up by singing to it, and we do this because we want it to awaken happy. Then everything will be well and good again. And when our world is going well then also we sing to the forest because we want it to share our happiness.'

All this I had heard before, but I had not realized quite so clearly that this was what the *molimo* was all about. It was as though the nightly chorus were an intimate communion between a people and their god, the forest. Moke even talked about this, only when he did so he stopped working on his bow and turned his wrinkled old face to stare at me with his deep, brown, smiling eyes. He told me how all pygmies have different

87

names for their god, but how they all know that it is really the same one. Just what it is, of course, they don't know, and that is why the name really does not matter very much. 'How can we know?' he asked. 'We can't see him, perhaps only when we die will we know and then we can't tell anyone. So how can we say what he is like or what his name is? But he must be good to give us so many things. He must be of the forest. So when we sing, we sing to the forest.'

The complete faith of the pygmies in the goodness of their forest world is perhaps best of all expressed in one of their great *molimo* songs, one of the songs that is only sung fully when someone has died. At no times do their songs ask for this or that to be done, for the hunt to be made better or for someone's illness to be cured . . . it is not necessary. All that is needful is to awaken the forest and everything will come right. But suppose it does not, suppose that someone died; then what? Then the men sit around their evening fire, as I had been doing with them for the past month, and they sing songs of devotion, songs of praise, to wake up the forest and rejoice it, to make it happy again. Of the disaster that has befallen them they sing, in this one great song, 'There is darkness all around us; but if darkness *is*, and the darkness is of the forest, then the darkness must be good.'

As Moke looked at me and spoke I wondered if this was how he had once explained things to his own son. Now I began to understand what it was that happened at night when fifteen feet of drainpipe was carried into the camp and someone blew or sang into it; and I began to understand what the 'presence' was—it was surely the presence of the forest itself, in all its beauty and goodness.

Moke turned back to his bow and began whittling again.

'You will soon see things of which you have never heard, and which you have never seen. Then you will understand things that I can never tell you. But you must stay awake—you may only see then once.'

He started humming to himself, and I left.

Chapter Five

EVERY night the men gathered to sing the songs of the *molimo*, and every morning the youths awakened the camp with their shouts and yells. This went on week after week, yet the daily life of the hunting group continued as though nothing particularly unusual was happening. The men were more sleepy in the mornings, and sometimes tempers were short; but never for long. The necessities of hunting and gathering for a living did not allow it.

As soon as the camp was light and there was no further danger of the *molimo* making an appearance, the women emerged from their huts and went down to the river to bathe and to collect water to cook the morning meal. Back in the camp they sat outside their huts, legs straight out in front of them in what seemed to be the most uncomfortable position possible. From there they prepared the family repast, placing green plantains in the hot ashes to roast or concocting a stew of mushrooms and chopped leaves with whatever meat was left over from the day before, covering the pot with large leaves to trap the steam and to keep out ashes and dirt. Any little chores that had to be done were done by their daughters who scurried around collecting dry twigs to heat the fire up more quickly, or who otherwise sat over the fire with their baby brothers, carefully guarding the pot and making sure it did not upset when the burning wood subsided. The younger women, freed in this way, relaxed and set to work making themselves beautiful. They sat in full view of the camp, unashamedly painting their bodies with black paste, calling their friends over to help with the more inaccessible regions. The buttocks in particular they found difficult to paint themselves, and rather than waste such an expansive area they would either bend over in front of a friend, or lie across her lap.

The menfolk combined business with pleasure too, usually meeting at the *kumamolimo* to drink from a steaming bowl of *liko*. The young bachelors mostly cooked their own simple breakfast here, and together they all discussed the day ahead of them, where they should go to hunt, what they might expect to find. Cephu never came over from his little camp at such times. He merely waited until everyone else set off for the hunt, then he joined them.

The men also had to prepare their nets. Helping each other, they uncoiled long lengths of knotted twine, sometimes stretching for three hundred feet, inspecting every section with care. Then, standing upright, the assistant holding the net off the ground, each man coiled his net again so that it hung from one shoulder to within a few inches of his ankle. With a loose length of twine he deftly bound the coils together, gave the net a final shake and hung it from some convenient branch or tree stump. Sometimes, when hunting, nets have to be set up rapidly, and anyone whose net becomes tangled through poor preparation and coiling can spoil the chances of all the others.

The women have little to do by way of preparation for the hunt, other than making sure of the strength of the bark tump line by which they carry their baskets. They put some food in these in case the day's hunt is a long one, and they wrap a glowing ember from the family's fire in large, damp leaves, and sit down and wait for the men to set off.

In normal times there is often a dance before everyone departs for the hunt. Men and women circle the camp, singing a hunting song, looking from right to left, clapping their hands and leaping extravagantly in imitation of the animals they hope to catch. But this was rare during the days of the *molimo* as we all felt the lack of rest.

Another form of hunting magic, if you can call it that, was a paste made from various parts of an antelope, particularly the heart and the eye. The charred flesh is mixed with spittle and ground to a paste, then put in an inverted antelope horn and stuck in the ground near the family fire. Just before going on the hunt, members of the family smear each other with some of the paste, using a twig to dig it out of the horn. It is difficult to say whether this form of magic is of pygmy or negro origin, but I rather think the latter because only a few families practised it, and they were highly criticized by the others as being antisocial. They were trying to get success for themselves at the expense of their friends. On one such occasion a family had a long run of good luck, the animals always falling into their net, while others had no luck at all. It was decided that this must have been due to *anjo*, as the medicine was called, so everyone agreed, including the offender, that the only thing to do was to destroy the horn that held the medicine. Everyone that had such a magic horn gave it to old Moke and promised they would not make any more selfish magic. Moke threw the horns in a fire and no more was said. It so happened that the families in question continued to have good luck for some time.

Hunting, for a pygmy group, is a co-operative affair—particularly net-hunting. The use of private magic is frowned on, though there is no law against it. The sacred hunting fire is all that they consider right and necessary, and this is found throughout the forest. It is thought to secure the blessing of the forest which provides the game, and to bring good luck to the entire group. Even those groups that hunt mainly with bow and arrow hunt not as individuals, but rather for the group as a whole. But for the net hunters it is impossible to hunt alone. Men, women and children all have to co-operate if the hunt is to be successful.

The 'Fire of the Hunt' is a simple act, involving the lighting of a fire at the base of a tree a short distance from the camp. In other pygmy groups I have seen a variation where the fire is lit within the camp, with special sticks around it, pointing in the direction the hunt is going to take. In this case the fire is surrounded by a long and heavy vine laid in a circle on the ground, and when the game is brought home it is placed within this circle before being divided. But this particular group, and others in the area, were less formal.

I remember one morning in particular when we went to kindle the Fire of the Hunt outside the camp, because that was the day that old Cephu committed one of the greatest sins possible in the forest.

I doubt if any of us had managed to snatch more than two hours' sleep, and we were all quiet while preparing for the hunt. In a pygmy camp this is the surest danger signal of all, for usually everyone is talking and laughing and shouting rude remarks from one end of the camp to the other. It was not only that we were tired, but of late Cephu had refused to contribute to the *molimo* basket, and that morning he had been heard to call out, in his loudest voice, that he was fed up with the *molimo* of 'that camp over there'. Even though he always made his camp a short distance off it was close enough to be thought of as the same camp, and whether or not we appreciated his presence we thought of his camp and ours being the same. Even his unwillingness to take part in the *molimo* was accepted, to maintain some semblance of unity; but this sudden statement made it impossible to ignore Cephu's feeling of rivalry any longer. Rather than cause any open breach, everyone in the main camp kept his thoughts to himself and was silent.

About an hour after dawn, four or five youths went off with their nets and spears to light the hunting fire. I went with them, and only when we had crossed the Lelo and were on the other side did they speak. Then it was about trivialities, and soon they were chatting and laughing as though nothing were wrong. Shortly afterwards we were joined by half a dozen of the younger married men, the most active hunters, and we lit the fire at the top of the hill. Leaves and twigs were piled around the base of a young tree, and an ember brought from the camp set them ablaze. More leaves were thrown on so that dense clouds of smoke went billowing up to the invisible sky, until finally the flames burst through with a great roar of victory. There was no great ritual or ceremonial, but somehow this act put the hunters in harmony with the forest and secured its blessing and assistance for the day's hunt. The pygmies regard fire as the most precious gift of the forest, and by offering it back to the forest they are acknowledging their debt and their dependence.

As we sat around waiting for the others, one or two couples passed by. They were going ahead so they would have extra time for gathering mushrooms on the way. They paused to chat and then walked on, lightly, swiftly, and gaily. Before long the main body of hunters arrived and asked where Cephu was. We had not seen him, and it seemed that he had left the camp shortly after us, but instead of passing by the hunting fire he had followed a different path. Someone suggested that he was building a fire of his own, and this brought cries of protest that not even Cephu would do such a thing. There was much shaking of heads, and when Ekianga arrived and was told what had happened he stood still for a moment, then turned around looking in all directions to see if there was any sign of smoke from another fire. He just said one word—'Cephu'—and spat on the ground.

By the time we go to the place where we were to make our first cast of the nets everyone was more silent and miserable than ever. It did not make matters any better to find Cephu there already, contentedly sitting over a fire eating roast plantains. He looked up and greeted us in a friendly way, and when questioned as to why he had followed his own path he opened his eyes wide and softly said that he had misunderstood and taken the wrong trail. There were angry remarks, but Cephu ignored them all and went on munching at his plantain, smiling at everyone in his kind, gentle way.

Ekianga and a few others made a brief reconnaissance and when they came back they gave instructions as to the best direction for setting up the nets. The womenfolk, who had been foraging around for mushrooms and nuts, picked up their baskets and went ahead with the small children. They trod lightly and made no noise beyond an occasional crunch of a rotten branch, buried deep beneath the carpet of leaves. We all spread out in a long semicircle, each man knowing exactly who should be to his right and who to his left. I went with Maipe, who had been sent by Njobo with his net. We soon lost sight and sound of the others, but Maipe knew just where he would be expected to set up his net, and he took a short cut. I lost all sense of direction and could not even tell on which side of us the women were lying in wait, listening for the signal to beat in towards our nets. We followed a little stream and paused by an enormous outcrop of huge stone boulders, almost perfectly square in shape, some of them eight feet across each face. Maipe looked around, then sat down to wait. After a few minutes Moke's nephew appeared to our left, stringing his net out through the undergrowth as he came.

The end of the net stopped a few feet short of the boulders, and Maipe deftly joined it to his net, then, slipping each coil off his shoulder in turn he hung his own net, fastening it to low branches and saplings. It stretched for some three hundred feet, so that one end was completely lost to view from the other. It stood about four feet high, and Maipe walked the length of it, silently adjusting it so that it touched the ground all the way along and was securely fastened above. If he found the net drooping where there was no support, he cut a sapling, stuck it in the ground and hung the net on that, bending the top sharply back, twisting it around and through the mesh so that the net could not slip. When this was done he took up his spear and casually sharpened it with a stone he picked off the ground.

It was about another five minutes before he suddenly stood up and beckoned me to do the same. He stood absolutely motionless, his head slightly on one side, listening; his spear was raised just a few inches from the ground. I listened too, but could hear nothing. The forest had become silent, even the crickets had stopped their almost incessant chirping. Maipe raised his spear higher, and then at some signal that I did not even notice there was a burst of shouting, yelling, hooting and clapping as the women and children started the beat. They must have been about half a mile away, and as they came closer the noise was deafening. We saw one antelope, a large red *sondu*, ears back, leaping towards the boulders as though it were heading straight for our net, but at the last moment it saw us and veered away to the left. Maipe could probably have killed it with his spear, but he said, 'That is not for us, it will probably fall into Ekianga's net.' Just then there was a lot of yelling from Moke's nephew. Maipe vaulted over the net and ran swiftly, leaping and bounding like the *sondu* to avoid obstacles. I followed as best I could, but was passed by several youngsters from further down the line before I reached the others. The *sondu* had gone into Ekianga's net, just as Maipe had said, but while all the attention was in that direction a water chevrotan, the *sindula*, had tried to fight its way through Moke's net.

The *sindula* is one of the most prized animals; not much larger than a small dog it is dangerous and vicious. Moke's nephew had been left all by himself to deal with it, as the others in that area were helping Ekianga with the *sondu*. The youngster, probably not much more than thirteen years old, had speared it with his first thrust, pinning it to the ground through the fleshy part of the stomach. But the animal was still very much alive, fighting for freedom. It had already bitten its way through the net, and now it was doubled up, gashing the spear shaft with its sharp teeth. Maipe put another spear into its neck, but it still writhed and fought. Not until a third spear pierced its heart did it give up the struggle.

It was at times like this that I found myself furthest removed from the pygmies. They stood around in an excited group, pointing at the dying animal and laughing. One boy, about nine years old, threw himself on the ground and curled up in a grotesque heap and imitated the *sindula's* last convulsions. The men pulled their spears out and joked with each other about being afraid of a little animal like that, and to emphasize his point one of them kicked the torn and bleeding body. Then Maipe's mother came and swept the blood-streaked animal up by its hind legs and threw it over her shoulder into the basket on her back.

At other times I have seen pygmies singeing feathers off birds that were still alive, explaining that the meat is more tender if death comes slowly. And the hunting dog, valuable as it is, gets kicked around mercilessly from the day it is born to the day it dies. I have never seen any attempt at the domestication of any animal or bird apart from the hunting dog. When I talked to the pygmies about it they laughed at me and said, 'The forest has given us animals for food, should we refuse this gift and starve?' I thought of turkey farms and Christmas, and of the millions of animals reared with the sole intention of slaughtering them for food in our own society.

Ekianga was busy cutting up the *sondu* by the time I reached his net, for it was too large an animal to fit in his wife's basket. Usually game is brought back to camp before it is divided, and in other groups the dead antelope would have been sent back immediately, around the sturdy neck and shoulders of one of the youngsters. But here the womenfolk crowded around as Ekianga hacked away, each claiming her share for her family. 'My husband lent you his spear . . .'; 'We gave your third wife some liver when she was hungry and you were away . . .'; 'My father and yours always hunted side by side . . .' these were all typical arguments, but for the most part they were not needed. Everyone knew who was entitled to a share, and by and large they stuck to the rules of the game.

Above all the clamour, and in the process of re-coiling the nets and assembling for the next cast, a disgruntled Cephu appeared and complained that he had had no luck. He looked enviously at the *sondu* and the *sindula*, but nobody offered him a share. Maipe's mother hurriedly covered the chevrotan with leaves to avoid argument. She worked efficiently, not bothering to take the basket off but working with her hands behind her back, throwing the dead animal up by bouncing the basket on her buttocks until it was completely hidden by the leaves.

We went on for a mile or two and made another cast. Once more Cephu was unlucky, and this time he complained even more loudly and accused the women of deliberately driving the animals away from his nets. They retorted that he had enough of his own womenfolk there, to which he replied rather un-gallantly, 'That makes no difference, they are a bunch of lazy empty-heads.'

They were still arguing when they set off for the third cast, but I had caught sight of old Moke, and stayed behind to talk to him. He had not left with the hunt, but had gone off on his own, as he usually did, with his bow and arrows. I was surprised to see him. He said, 'Don't follow them any longer, they will deafen you with their noise. Cephu will spoil the hunt completely—you see.' He added that he had happened to be near-by, waiting for the animals either to break through the nets or to escape around the edges. He picked a large civet cat from the ground, the skin slightly stained where it had been pierced by an arrow. 'Not very good for eating,' Moke commented, 'but it will make a beautiful hat. Cephu won't get even that— he is too busy watching other people's nets to watch his own. His is a good net to stand behind with bow and arrow!' He chuckled to himself and swung the cat high in the air.

We ambled back slowly, Moke talking away, sometimes to himself and sometimes to me. On the way he stopped to examine some tracks that were fresh, appearing over the tracks made by the hunters earlier on. He announced that it was a large male leopard, and that it was probably watching us from 'over there' . . . waving casually to our right with his bow. Then he moved on in his leisurely way, humming to himself.

Back in the camp I was surprised to find that some of the hunters were back already, including Maipe. They had taken a short cut, and travelled twice as fast as Moke and myself. Some of them said it was because rain had threatened, but others, more bluntly, said it was because they did not like the noise Cephu was making. There were a number of women in the camp, and they seemed anxious to change the subject, so Moke told them about the leopard tracks. They laughed and said what a shock the hunters might get if they came back along that trail. One of them started miming the leopard lying in wait, its eyes staring from this side to that. The others formed a line and pretended to be the hunters. Every few steps the 'leopard' turned around and jumped up in the air, growling fiercely, sending its pursuers flying to the protection of the trees.

This dance was still in progress when the main body of the hunters returned. They strode into camp with glowering faces and threw their nets on the ground outside their huts. Then they sat down, with their chins in their hands, staring into space and saying nothing. The women followed, mostly with empty baskets, but they were by no means silent. They swore at each other, they swore at their husbands, and most of all they swore at Cephu. Moke looked across at me and smiled. He was skinning the civet.

I tried to find out what had happened, but nobody would say. Kenge, who had been sleeping, came out of our hut and joined the shouting. He was the only male who was not sitting down, and although he was young he had a powerful voice, and a colourful usage of language. I heard him saying, 'Cephu is an impotent old fool. No he isn't, he is an impotent old animal—we have treated him like a man for long enough, now we should treat him like an animal. *ANIMAL!*' He shouted the final epithet across at Cephu's camp, although Cephu had not yet returned.

The result of Kenge's tirade was that everyone calmed down and began criticizing Cephu a little less heatedly, but on every possible score. The way he always built his camp separately, the way he had even referred to it as a separate camp, the way he mistreated his relatives, his general deceitfulness, the dirtiness of his camp, and even his own personal habits.

The rest of the hunters came in shortly afterwards, with Cephu leading. He strode across the camp and into his own little clearing without a word. Ekianga and Manyalibo, who brought up the rear, sat down at the *kumamolimo* and announced to the world at large that Cephu had disgraced them all, and that they were going to tear down the *kumamolimo* and abandon the camp and abandon the *molimo*. Manyalibo shouted that he wanted everyone to come to the *kumamolimo* at once, even Cephu. This was a great matter and had to be settled immediately. Kenge, not slow to seize the opportunity for reviving his earlier show of wit relayed Manyalibo's message by standing in mid-camp and shouting over in Cephu's direction, '*Nyama'e, Nyama'e; pika'i to, Nyama'ue!*' . . . 'Animal there! Animal there! Come at once, you, you animal!' The youngsters all laughed loudly, but the older men paid no attention. This sent Kenge into a sulk and he started calling us all animals, adding that we were a lot of savages, just like the village people. But now not even the youngsters were listening, for Cephu had appeared.

Trying not to walk too quicky, yet afraid to dawdle too deliberately, he made an awkward entrance. For as good an actor as Cephu it was surprising. By the time he got to the *kumamolimo* everyone was doing something to occupy himself. They were staring into the fire or up at the tree-tops; roasting plantains, smoking, or whittling away at arrow shafts. Only Ekianga and Manyalibo looked impatient, but they said nothing. Cephu walked into the group, and still nobody spoke. He went up to where a youth was sitting in a chair. Usually he would have been offered a seat without his having to ask, and now he did not dare ask, and the youth continued sitting down in as nonchalant a manner as he could muster. Cephu went to another chair where Amabosu was sitting. He shook it violently when Amabosu ignored him, at which he was told, 'Animals lie on the ground.'

This was too much for Cephu, and he went into a long tirade about how he was one of the oldest hunters in the group, and one of the best hunters, and that he thought it was very wrong for everyone to treat him like an animal. Only villagers were like animals. Masisi spoke up in his favour, and ordered Amabosu to give up his chair. Amabosu did so with a gesture of contempt, and moved over to the far side of the group. Moke was the only one to stay completely apart. He sat outside his hut. He had finished skinning the civet, and was already busy twisting strands of vine together to make a frame for the hat. He looked peaceful and contented.

Manyalibo stood up and began a rather pompous statement of how everyone wanted this camp to be a good camp, and how everyone wanted the *molimo* to be a good *molimo* with lots of singing, lots of eating, and lots of smoking. But Cephu never took part in the *molimo*, he pointed out. And Cephu's little group never contributed to the *molimo* basket. Manyalibo went on, listing the various ways in which Cephu had defaulted and made the camp into a bad camp instead of a good camp, and the *molimo* a bad *molimo*. And now, he added, Cephu had made the hunt a bad hunt. He recalled how on the first day, even before the camp was built, we had been presented with an antelope by the forest, but how ever since Cephu had joined us things had gone from bad to worse.

Cephu tried to interject here and there that the *molimo* was really none of his business, at which Masisi, who had befriended him over the chair incident, and who had relatives in Cephu's camp, rounded on him sharply. He reminded Cephu that he had been glad enough to accept help and food and song when his daughter had died; now that his 'mother' had died why did he reject her? Cephu replied that Balekimito was not his mother. This was what everyone was waiting for. Not only did he name the dead woman, an unheard-of offence, but he denied that she was his mother. Even though there was only the most distant relationship, and that by marriage, it was equivalent to asserting that he did not belong to the same group as Ekianga and Manyalibo and the rest of us, and Ekianga leapt to his feet and brandished his hairy fist across the fire. He said that he hoped Cephu would fall on his spear and kill himself like the animal he was. Who else but an animal would steal meat from others? There were cries of rage from everyone, and Cephu burst into tears.

99

Apparently, during the last cast of the nets Cephu, who had not trapped a single animal the whole day long, had slipped away from the others and set up his net in front of them. In this way he caught the first of the animals fleeing from the beaters, but he had not been able to retreat before he was discovered.

I had never heard of this happening before, and it was obviously a serious offence. In a small and tightly knit hunting band survival can only be achieved by the closest co-operation and by an elaborate system of reciprocal obligations which ensures that everyone has some share in the day's catch. Some days one gets more than others, but nobody ever goes without. There is, as often as not, a great deal of squabbling over the division of the game, but that is expected, and nobody attempts to take what is not his due.

Cephu tried very weakly to say that he had lost touch with the others and was still waiting when he heard the beating begin. It was only then that he had set up his net, where he was. Knowing that nobody believed him, he added that in any case he felt he deserved a better place in the line of nets. After all, was he not an important man, a Chief, in fact, of his own band? Manyalibo tugged at Ekianga to sit down, and sitting down himself he said there was obviously no use prolonging the discussion. Cephu was a big Chief, and a Chief was a villager, for the BaMbuti never have Chiefs. And Cephu had his own band, of which he was Chief, so let him go with it and hunt elsewhere and be a Chief elsewhere. Manyalibo ended a very eloquent speech with, 'Pisa me taba . . .' 'Pass me the tobacco.'

Cephu knew he was defeated and humiliated. Alone, his band of three or four families was too small to make an efficient hunting unit. He apologized profusely, reiterating that he really did not know he had set up his nets in front of the others, and that in any case he would hand over all the meat. This settled the matter, and accompanied by most of the group he returned to his little camp and brusquely ordered his wife to hand over the spoils. She had little chance to refuse, as hands were already reaching into her basket and under the leaves of the roof of her hut where she had hidden some liver in anticipation of just such a contingency. Even her cooking pot was emptied. Then each of the other huts was searched and all the meat taken. Cephu's family protested loudly and Cephu tried hard to cry, but this time it was forced and everyone laughed at him. He clutched his stomach and said he would die; die because he was hungry and his brothers had taken away all his food; die because he was not respected.

The *kumamolimo* was festive once again, and the camp seemed restored to good spirits. An hour later, when it was dark and the fires were flickering outside every hut, there was a great blaze at the central hearth and the men talked about the morrow's hunt. From Cephu's camp came the sound of the old man, still trying hard to cry, moaning about his unfortunate situation, making noises that were meant to indicate hunger. From our own camp came the jeers of women, ridiculing him and imitating his moans.

When Masisi had finished his meal he took a pot full of meat cooked by his wife, with mushroom sauce, and quietly slipped away into the shadows in the direction of his unhappy kinsman. The moaning stopped, and when the evening *molimo* singing was at its height I saw Cephu in our midst. Like most of us he was sitting on the ground in the manner of an animal. But he was singing, and that meant that he was just as much a MuMbuti as anyone else.

Chapter Six

CEPHU had committed what is probably one of the most heinous crimes in pygmy eyes, and one that rarely occurs. Yet it was settled simply and effectively, without any evident legal system being brought into force. It can not be said that Cephu went unpunished though, because for those few hours when nobody would speak to him he must have suffered the equivalent of as many days solitary confinement for us. To have been refused a chair by a mere youth, not even one of the great hunters; to have been laughed at by women and children; to have been ignored by men; all these things would not be quickly forgotten. Without any formal process of law Cephu had been firmly put in his place, and it was unlikely he would do the same thing again in a hurry.

So it was with all pygmy life, on the surface at least. There was a confusing, seductive informality about everything they did. Whether it was a birth, a wedding, or a funeral, in a pygmy hunting camp or in a negro village, there is always an unexpectedly casual, almost carefree attitude. There was, for instance, no apparent specialization; everyone took part in everything. Children had little or no say in adult affairs, but the only rigid exclusion from adult activities seemed to be with regard to certain songs, and of course the *molimo*. Between men and women there was also a certain degree of specialization, but little that could be called exclusive.

There were no chiefs, no formal councils. In each aspect of pygmy life there might be one or two men or women who were more prominent than others, but usually for good, practical reasons. This showed up most clearly of all in the settling of disputes. There was no judge, no jury, no court. The negro tribes all around had their tribunals, but not so the pygmies. Each dispute was settled as it arose, according to its nature.

Roughly there were four ways of settling a dispute, each way acting as an efficient deterrent but without necessitating any system of outright punishment. In a small and co-operative group no individual would want the job either of passing judgment or of administering punishment, so like everything else in pygmy life the maintenance of law was a co-operative affair. Certain offences, rarely committed, were considered so terrible that they would of themselves bring some form of supernatural retribution. Others became the affair of the *molimo*, which in its morning rampages showed public disapproval by attacking the hut of the culprit, possibly the culprit himself. Both these types of crime were extremely rare. The more serious of the other crimes, such as theft, were dealt with by a sound thrashing administered co-operatively by anyone who felt inclined to participate, but only after the entire camp had been involved in discussing the case. Less serious offences were settled in the most simple way, by the litigants either arguing it out between themselves, or engaging in a mild fight.

I only came across one instance of the first type of crime. We had all eaten in the evening and were sitting around our fires. I was with Kenge and a group of the bachelors, talking about something quite trivial, when all of a sudden there was a tremendous wailing and crying from Cephu's camp. A few seconds later there was shouting from the path connecting the two camps and young Kelemoke came rushing through our camp, hotly pursued by other youths who were armed with spears and knives. Everyone ran into their huts and closed the little doorways. The bachelors, however, instead of going in with their families, ran to the nearest trees and climbed up into the lower branches. I followed Kenge up one of the smaller trees and sat amongst the ants whose persistent biting I hardly noticed. What was going on below was much more interesting.

Kelemoke tried to take refuge in a hut, but he was turned away with angry remarks, and a burning log was thrown after him. Masisi yelled at him to run to the forest. His pursuers were nearly on top of him when they all disappeared at the far end. At this point three girls came running out of Cephu's camp, right into the middle of our clearing. They also carried knives—the little paring knives they use for cutting vines and for peeling and shredding roots. They were not only shouting curses against Kelemoke and his immediate family, but they were weeping, with tears streaming down their faces. Not

103

finding Kelemoke one of them threw her knife on to the ground and started beating herself with her fists, shouting: 'He has killed me, he has killed me!' over and over again. After a pause for breath she added, 'I shall never be able to live again!' Kenge made a caustic comment on the logic of the statement, from the safety of his branch, and immediately the girls turned their attention to our tree and began to threaten us. They called us all manner of names, and then they fell on the ground and rolled over and over, beating themselves, tearing their hair, and wailing loudly.

Just then there were more shouts from two directions, one from the youths who had evidently found Kelemoke hiding just outside the camp. At this the girls leap up, and brandishing their knives they set off to join the pursuit. The other shouts, for the first time of adult men and women, came from the camp of Cephu. I could not make out what they were about, but I could see the glow of flames.

I asked Kenge what had happened. He looked very grave now, and said that it was the greatest shame that could befall a pygmy; Kelemoke had committed incest. In some African tribes it is actually preferred that cousins should marry each other, but amongst the BaMbuti this was considered almost as incestuous as sleeping with a brother or sister. I asked Kenge if they would kill Kelemoke if they found him, but Kenge said they would not find him. 'He has been driven to the forest,' he said, 'and he will have to live there alone. Nobody will accept him into their group after what he has done. And he will die, because one cannot live alone in the forest. The forest will kill him. And if it does not kill him he will die of leprosy.' Then, in typical pygmy fashion, he burst into smothered laughter, clapped his hands and said, 'He has been doing it for months; he must have been very stupid to let himself be caught. No wonder they chased him into the forest.' To Kenge, evidently, the greater crime was Kelemoke's stupidity in being found out!

All the doors to the huts below still remained tight shut. Njobo and Moke had both been called on by the youths to show themselves and settle the matter, but they had both refused to leave their huts, leaving Kelemoke to his fate. Now there was an even greater uproar over in Cephu's camp, and

Kenge and I decided to climb down the tree and see what was
going on.

One of the huts was in flames, and people were standing all
around either crying or shouting. There was a lot of struggling
going on amongst a small group of men, and women were
brandishing fists in each other's faces. We decided to go back
to the main camp, and by then it was filled with clusters of
men and women, standing about discussing the affair. Not long
afterwards a contingent from Cephu's group came over and
demanded a discussion. They swore at the children who, delighted by the whole thing, were imitating the epic flight and
pursuit of Kelemoke. The adults, however, were in no joking
frame of mind, and sat down to have a discussion. But the talking did not centre around Kelemoke's act so much as around
the burning of the hut. Almost everyone seemed to dismiss the
act of incest. Masalito, one of Kelemoke's uncles, and with
whom he had been staying since his father died, had great
tears rolling down his cheeks, and he said, 'Kelemoke only did
what any youth would do, and he has been caught and driven
to the forest. The forest will kill him. That is finished. But my
own brother has burned down my hut and I have nowhere to
sleep, I shall be cold. And what if it rains? I shall die of cold
and wet at the hands of my brother.'

The brother, Aberi, instead of taking up the point that
Kelemoke had dishonoured his daughter, made a feeble pro-
testation that he had been insulted, Kelemoke should have
taken more care. And Masalito should have taught him better.
This started the argument off on different lines, and both families fought bitterly, accusing each other for more than an hour.
Then the elders began to yawn and say they were tired, they
wanted to go to sleep, we would settle it the next day.

For a long time that night the camp was alive with whispered remarks, and not a few rude jokes were thrown about
from one hut to the other. The next day I went over to Cephu's
camp and found the girl's mother busy helping to rebuild
Masalito's hut; the two men were sitting down and talking as
though nothing much had happened. All the youths told me
not to worry about Kelemoke, that they were secretly bringing
him food in the forest, he was not far away. Three days later,

when the hunt returned in the late afternoon, Kelemoke came wandering idly into the camp behind them, as though he too had been hunting. He looked around cautiously, but nobody said a word or even looked at him. If they ignored him, at least they did not curse him. He came over to the bachelor's fire and sat down. For several minutes the conversation continued as though he were not there, and I saw his face twitching. But he was too proud to speak first. Then a small child was sent over by her mother with a bowl of food, which she put in Kelemoke's hands and gave him a shy, friendly smile.

Kelemoke never flirted with his cousin again, and now, five years later, he is happily married and has two fine children. He does not have leprosy, and he is one of the best liked and most respected of the hunters.

*   *   *

I have never heard of anyone being completely ostracized, but the threat always is there, and that is sufficient to ensure good behaviour. The two attitudes the pygmy dislikes most are contempt and ridicule. Contempt is most effectively shown by ignoring someone, even if he comes and sits at the same fire. But it is difficult to maintain for long because as a hunter you cannot afford to ignore a fellow hunter for long. So ostracism of this kind is infrequent and does not last for long. Almost as effective, and of much longer duration, is ridicule. The BaMbuti are good-natured people with an irresistible sense of humour; they are always making jokes about each other, even about themselves, but their humour can be turned into an instrument of punishment when they want.

Such a case was a great fight between the two same brothers who had fallen out over Kelemoke's misdemeanours, Aberi and Masalito. This incident shows as well as any how a tiny and often imagined insult can be magnified out of all proportion until it becomes a serious issue, threatening the peace of the camp. In any small, closed community, there are hidden tensions everlastingly at play. They are capricious, they are fluctuating and vapid, but everyone of them is potentially explosive.

Of the two brothers Aberi was the elder. He was aggressive

and ugly, and he considered himself the head of the family. The oldest brother, Kelemoke's father, had died. But Kelemoke had been brought up by Masalito rather than Aberi because everyone agreed that Masalito, if younger, had much more sense. He was also a better hunter, a better singer, a better dancer, and considerably better looking. He had a prettier wife too, for Aberi's wife had lost half her face, eaten away by yaws. So between these two brothers there were always strained feelings. Open conflict was avoided because Masalito usually lived with the main group, of which his wife was a member, while Aberi always hunted with Cephu, his cousin.

But Masalito was kindly and liked peace and quiet, so when Cephu joined the main group at Apa Lelo he often used to visit his brother in Cephu's camp. Aberi's wife was a hard-working woman, and although it was difficult to look at her without flinching she was loved by everyone. One afternoon, on a rainy day, Masalito wandered over to visit his brother. Rainy days are not happy days in a hunting camp; they are often oppressive, and of course hunting is impossible. Everyone is cooped up in their tiny leaf hut, thinking of all the game that is going uncaught. Aberi was sleeping, so Masalito sat down beside his sister-in-law, Tamasa, and asked her for a smoke. She passed him the clay pipe bowl and a little tobacco, and a long stem cut freshly from the centre of a plantain leaf. She had recently come back from a visit to a negro village and she had brought several of these stems with her, as they give by far the best smoke.

Masalito smoked a few puffs, then passed the pipe to Tamasa. But instead of accepting it she abruptly turned it upside down and knocked the remaining tobacco out into the dirt. This was not only wasteful, but a deliberate insult. Masalito did not want to take offence, so he gave Tamasa the chance to re-establish good relations. He asked her to give him one of the new pipe stems to take back to his hut for himself and his wife to enjoy. To refuse her husband's brother would be an open act of aggression quite uncalled for, and if she made the gift she would automatically cancel the previous insult. Masalito thought that he had not only manoeuvred the political situation to his advantage but that he had also won for himself and his

THE FOREST PEOPLE

wife a new pipe stem. Unfortunately he had chosen a bad day.
Tamasa felt that Masalito was trying to get the better of her,
and was trying to get her to give away her husband's property
while he was asleep, instead of asking him. So she said, 'Certainly I will give you a pipe stem.' She went behind the hut,
where all the rubbish was thrown, and she came back with a
withered and smelly stump of a pipe that had been pared away
so often between smokes, to keep the mouthpiece fresh, that it
was now only two feet long instead of six or seven. And instead
of being fresh and green and juicy it was stale and brown and
dry, reeking of old tobacco and spittle. This she presented to
Masalito who was so upset that he cried with rage and asked
why she and Aberi always treated him so badly. Tamasa
promptly retorted that it was not bad treatment to feed him
and give him tobacco every time he came over from his camp.
Didn't his wife look after him, that he had to come and scrounge
from his brother?

This was too much for Masalito, who had only wanted to be
on good terms with everyone. He called Tamasa a name that
one should never use to one's brother's wife, and Tamasa
started crying. This awakened Aberi who came out of his hut
sleepily, looking more ugly than ever. Without trying to explain, Masalito, still sobbing tearlessly with rage, hit his
brother over the head with the withered pipe stem Tamasa had
given him. Brandishing the offending article above his head he
stalked back to the main camp shouting at the top of his voice
what miserable wretches his brother and his sister-in-law were.
The dirty old pipe was damning evidence, and all sympathy
was with Masalito. Njobo's wife took her husband's pipe stem
and gave it to Masalito, who quickly regained his good humour.
But by now Aberi was aroused, and came storming into the
camp saying that his brother had insulted Tamasa and that it
was a matter for serious action. Masalito should be thrashed.

Everybody thought this was a great joke, and burst into
laughter. Only children and youths get thrashed, and Masalito
was a father. Aberi did not like being laughed at and he went
over to where Masalito was sitting down on a log, smoking his
new pipe. He demanded an apology, but Masalito, now completely composed, simply picked up the old pipe stem and

threw it at his brother's feet, saying, 'There! That is for her!'
By now a crowd had gathered around, and the hotter Aberi
became the more relaxed Masalito seemed to be.

Aberi shook his fists and said that if nobody else was going
to thrash Masalito he was. This brought more laughs as Aberi
was positively puny beside Masalito, who himself was little
over four feet tall. Masalito invited his brother to come and try.
Aberi then pretended that he was holding a spear and in a
high-pitched voice, squeaking with anger, he cried out, 'I am
going to get my spear and I am going to kill you completely!'
with which he imitated a spear thrust in Masalito's direction.
There was a gasp of horror as this is something one does not
say even in jest to one's brother. But Masalito replied calmly,
'Go and get your spear, then, and come back and kill me. I'll
still be here. You don't have the courage to kill your brother.'
He said a lot of other things, goading Aberi on to an even higher
pitch of fury. Aberi tried to make himself more impressive by a
graphic dance, which was meant to show exactly how he was
going to leap in the air and twist around and drive the spear
home. But he was not a good dancer and when he tried to illustrate the leap he fell flat on his face.

That was the end of the matter for Aberi. For weeks he was
ridiculed, everyone asking him if he had lost his spear, or telling
him to be careful not to trip and fall. But it was not quite the
end of the matter for Masalito. He felt more angry than ever
at his brother, because Aberi had made himself, and so the
whole family, an object of ridicule. Masalito felt this an added
insult to himself. Even Kelemoke was furious. So relations between the two brothers got worse and worse, and every time
there was a silence in the camp Masalito would complain in a
loud voice about his brother and his brother's wife. At first this
was tolerated, but finally it threatened to split the camp in two
as Masalito began naming all his friends and relatives to support him in ignoring Aberi; he himself refused to say a word to
his brother. Of course his brother could claim support from the
same relatives and friends, and tension was spreading throughout both camps. That was when the *molimo* stepped in and
showed public disapproval of Masalito. The original source of
the dispute, and Aberi's undoubted wrongdoing, was forgotten.

Masalito was guilty of the much more serious crime of splitting the hunting band into opposing factions. It ended by he himself being ridiculed and shamed into silence. After that it was only one or two days before the two brothers became good friends and hunting companions once again.

*   *   *

Disputes are generally settled with little reference to the alleged rights and wrongs of the case, but more with the sole intention of restoring peace to the community. One night Kenge slipped out of our hut on an amorous expedition to the hut of Manyalibo, father of an attractive maiden who was one of Kenge's many admirers. Shortly afterwards there was a howl of rage and Kenge came flying back across the clearing with a furious Manyalibo hurling sticks and stones after him. Manyalibo then took up a position in the middle of the clearing and woke the whole camp up, calling out in a loud voice and denouncing Kenge as an incestuous good-for-nothing. Actually Kenge was only very distantly related to the girl, and the flirtation was not at all out of order, though marriage might have been. Several people tried to point this out to Manyalibo, but he became increasingly vociferous. He said that it wasn't so much that Kenge had tried to sleep with his daughter, but that Kenge had been brazen enough to crawl right over her sleeping father to get at her, waking him up in the process. This was a considered insult, for any decent youth would have made a prior arrangement to meet his girl elsewhere. He called on Kenge to justify himself. But Kenge was too busy laughing, and only managed to call out, 'You are making too much noise!' This seemed a poor defence, but in fact it was not. Manyalibo set up another hue and cry about Kenge's general immorality and disrespect for his elders, and strode up and down the camp rattling on the roofs of huts to call everyone to his defence. But Moke took his place in the centre of the camp, where Manyalibo had stood, and where everyone stands who wants to address the whole camp formally, and he gave a low whistle, like the whistle given on the hunt to call for silence. When everyone was quiet he told Manyalibo that the noise was giving him a headache, and he wanted to sleep. Manyalibo retorted

that this matter was more serious than Moke's sleep, to which
Moke replied in a very deliberate, quiet voice: 'You are
making too much noise . . . you are killing the forest, you
are killing the hunt. It is for us older men to sleep at night and
not to worry about the youngsters. They know what to do and
what not to do.' Manyalibo growled with dissatisfaction, but
went back to his hut, taunted by well-directed remarks from
Kenge and his friends.

Whether Kenge had done something wrong or not was relatively immaterial. Manyalibo had done the greater wrong by
waking the whole camp and by making so much noise that all
the animals would be frightened away, spoiling the next day's
hunting. The pygmies have a saying that a noisy camp is a
hungry camp.

*　　　*　　　*

Some misdemeanours are very simply dealt with. A pygmy
thinks nothing of stealing from negroes; they are, after all,
only animals . . . as seen by pygmy eyes. But amongst themselves theft is virtually non-existent. For one thing they have
few possessions, and for another there is no necessity for theft
except through laziness. Pepei was such a lazy pygmy, and he
was always stealing from the negroes. But out in the forest he
was forced to work for his food, to build his own hut if he
wanted one (he was a bachelor but liked having his own hut,
sharing it with his younger brother), and he had to make his
own bows and arrows and hunting net, and to cook his own food.

This was hard on Pepei, and although he meant no harm
he could not resist slipping around the camp at night, taking
a leaf from this hut and a leaf from that, until he had enough
to thatch his own hut—which he made by acquiring saplings in
a similar manner. Food used to disappear mysteriously, and
Pepei had always seen a dog stealing it. But finally he was
caught in the act by old Sau, Amabosu's mother, who was a
very strange and frightening old lady. Pepei crept into her hut
one night and was lifting the lid of a pot when she smacked him
hard on the wrist with a wooden pestle. She then grabbed Pepei
by the arm, twisting it behind his back, and forced him out into
the open.

Nobody really minded Pepei stealing because he was a born comic and a great story-teller. But he had gone too far in stealing from old Sau, who had lost her husband and was supported by her son, Amabosu, and his wife. So the men ran out of their huts angrily and caught hold of Pepei while the youths broke off thorny branches and whipped him until he managed to break away. He went running, as fast as he could, into the forest; he cried bitterly and wrapped his arms around himself for comfort. He stayed in the forest for nearly twenty-four hours, and when he came back the next night he went straight to his hut, unseen, and lay down to sleep. His hut was between mine and Sau's, and I heard him come in, and I heard him crying softly because even his brother wouldn't speak to him.

The next day Pepei was his old self, and everyone was glad to see him laughing; they were happy to be able to listen to his jokes, and they all gave him so much food that he wouldn't have to steal again.

*   *   *

But sometimes a dispute cannot be settled in any of these ways. It blows up too quickly to be ended by ridicule; the participants are too old to be thrashed; yet it is not serious enough to merit ostracism. Like most disputes they are usually trivial in origin, and often arise from the confined and close conditions of living. Personal relations become very involved in a hunting group, particularly when both members of a sister-exchange marriage are living in the same group.

Under this system, when a boy chooses his wife he becomes obliged to find a 'sister' (in fact any related girl) to offer in exchange to his bride's family for one of their bachelor sons. This can be quite a chore, as it may be difficult to find a sister who is willing to marry whoever his in-laws have in mind as a groom, and whom the groom himself will also like. Amabosu had married Ekianga's sister, and Ekianga had married, as his third wife, Amabosu's sister, the beautiful Kamaikan. He had just had a child by her, and Amabosu was still childless. But having had a child by her he was under an obligation not to sleep with her and his two other wives were happy that at last they could expect him to pay them some attention. But to everyone's dismay he continued sleeping with Kamaikan. And one day Amabosu started making loud remarks across the camp about his sister's health.

This was not at Apa Lelo, but at a later camp called Apa Kadi-ketu. Amabosu's hut was to my right, and it faced directly across to the first of Ekianga's huts; the largest and the best, built by himself and Kamaikan. Amabosu was sitting down at his fire at the time, and his wife was sitting near him, peeling mushrooms for the evening meal. Ekianga was inside the hut with Kamaikan who had given the baby to her old mother, Sau, to look after for a few hours. At first, Amabosu's remarks brought no response from the closed hut. But he became more and more explicit, denouncing Ekianga's shame in considerable and intimate detail, hinting that what was going on inside the hut at that moment was precisely what every recent mother should avoid if she was to be able to look after the baby properly and give it lots of milk.

At last Ekianga retorted angrily from inside the hut. Amabosu continued sitting quietly, staring with his wide, strange eyes into the fire, and he answered in as caustic tones as he could find. In the end, Ekianga's door burst open and Ekianga appeared, spear in hand. He looked wild with fury and shame, and he flexed his muscles and hurled the spear at Amabosu with all his might. Amabosu looked up from the fire for the first time, and sat perfectly still with not a flicker of emotion on his face, his eyes still wide and cold and staring. The spear struck the ground within twelve inches of his feet, which was probably exactly where it had been intended to land. Amabosu calmly rose, pulled the spear free, and tossed it behind his hut into the rubbish dump. This was just about as insulting a gesture as he could have made at that moment, and it started one of the most dramatic fights I have seen in a pygmy camp, and one of the most complicated.

As long as it was between the two men all was well, but the wives were torn between dual loyalties as wives and sisters. Amabosu's wife entered the fray by swearing at her husband for throwing her brother's spear into the rubbish dump. Amabosu countered by smacking her firmly across the face. Normally Ekianga would have approved of such manly assertion of authority over a disloyal wife, but as the wife was his sister he countered by going into his hut and dragging out Kamaikan, whom he in turn publicly smacked across her face. But Kamaikan was made of tough stock. She picked up a log from the fire and beat her husband over the back with it. It is difficult to remember just exactly what the sequences of events was after that, but my notes contain the following incidents.

The two wives began fighting tooth and nail quite literally. The men battled with burning logs, three or four feet long, swinging them from the cool end, but always just missing each other. Every now and again one of the girls would break free and go to the assistance of her brother, and the other would run and pull her away.

As always, a large crowd gathered—men, women and children. But nobody volunteered to stop the fight or to adjudicate in any way. They took one side or the other, either according to the rights and wrongs of the issue, or merely on the grounds that one or the other was the better fighter and most likely to win. The general opinion was that Ekianga should certainly not be sleeping with Kamaikan while she was nursing a baby, though it was Kamaikan's fault as well so Amabosu was wrong in defending her. But by and large the onlookers were more interested in the fight than the issue at stake.

As the fight grew more serious, and some visible damage was being done to all four parties, first relatives, and then friends, began to take sides more vigorously, and it looked as though fighting was going to break out all over the camp. At that moment Arobanai, Ekianga's senior wife, came striding out of her hut and announced in her deep contralto voice that she couldn't rest with all this noise going on, so it would have to stop. To make her impartiality quite plain she added that her animal of a husband could sleep with anyone he wanted to as far as she was concerned. She then went to where the two girls were locked together, both looking very much the worse for wear. She pulled them apart, and having given Amabosu's wife a hefty kick she took Kamaikan tenderly and led her back to her own hut.

At this old Sau, who had been sitting on the ground throughout, with Kamaikan's baby on her lap, handed the baby to her weeping daughter-in-law and slowly got to her feet. Her son and Ekianga were now each armed with stout sticks, about four inches thick and three feet long, and they were lashing out with no reservations. Sau hobbled up to them and calmly pushed her way between, then she turned and put her head against Ekianga's stomach and her backside against her son's legs, and forced them apart. They continued trying to fight each other around the old lady, reaching out far so as not to hit her, but it was just too difficult. Ekianga was the first to throw away his weapon, which he did over Sau's head, with all his might. But Amabosu ducked and it missed him. Amabosu then threw his club on the ground and led his mother back to where his wife had been joined by Kamaikan, both of whom were playing happily with the baby.

*   *   *

This is perhaps one of the most remarkable features of pygmy life, the way everything settles itself with apparent lack of organization. Co-operation is the key to pygmy society; you can expect it and you can demand it and you have to give it. If your wife nags you at night so that you cannot sleep you merely have to raise your voice and call on your friends and relatives to help you. Your wife will do the same, so whether you like it or not the whole camp becomes involved. It is at this point that someone, very often an older man or woman who has so many relatives and friends that he cannot be accused of being partisan, steps in with the familiar remark that everyone is making too much noise, or else he will divert the issue on to a totally different tack so that people forget the origin of the argument and give it up.

*   *   *

Issues other than disputes are settled the same way, without leadership appearing from any particular individual. If it is a question of the hunt, every adult male discusses it until there is agreement. The women can throw in their opinions, particularly if they know that the area the men have selected is barren of vegetable foods. But the men usually know this anyway. If the issue is one of marriage, and a father announces that he does not like the girl his son has chosen, his son can call on all his friends to help him. If he is strong and holds out, then the whole group will be assembled to discuss the case. If they agree with the father, then either the boy has to give up his talk of matrimony or else make up his mind that he is going to marry her anyway. In the latter case he would probably go and live with her hunting group. But it is seldom that things come to such a head.

In fact pygmies dislike and avoid personal authority, though they are by no means devoid of a sense of responsibility. It is rather that they think of responsibility as a communal factor. If you ask a father or a husband why he allows his son to flirt with a married girl or his wife to flirt with other men, he will answer, 'It is not my affair,' and he is right. It is *their* affair, and the affair of the other men and women, and of their brothers and sisters. He will try to settle it himself, either by argument or by a good beating, but if this fails he brings everyone else into the dispute so that he is absolved of personal responsibility.

If you ask a pygmy why he has no chiefs, no law-givers, no councils, or no leaders, he will answer with misleading simplicity, 'Because we are the People of the Forest.' The forest, the great provider, is the one standard by which all deeds and thoughts are judged; it is the chief, the law-giver, the leader, and the final arbitrator.

Chapter Seven

IN THE forest life appears to be free and easy, happy-go-lucky, with a certain amount of perpetual disorder as a result. But in fact, beneath it all, there is order and reason; reaching everywhere is the firm, controlling hand of the forest itself.

If individual responsibility and authority are shunned and dispersed in other aspects of pygmy life, as in the maintenance of law and order, it is no less so with regard to the upbringing of children. It is no accident that you call everyone who is of the same age group as your parents either 'father' or 'mother'; those still older you call 'grandparent'; those of the same age as yourself you refer to by a term which could be translated as either 'brother' or 'sister', and anyone younger is 'child'. More often, especially when addressing anyone of the same age or younger, you just call them by name. In this way a pygmy camp is much less formal than most African societies, where there are not only different terms of address and reference for different degrees of relationship, but also correspondingly different modes of behaviour that are enforced at all times.

During the day one pygmy camp looks much like another. The sun filters down through the trees and brightens the camp with shafts of spiralling light as it catches the columns of smoke drifting lazily upwards. This is what makes a camp look so full of life at midday, even when most people are out on the hunt and those that have stayed behind are dozing in the shade of their little leaf huts. The light has a life of its own, and after it has danced down the coils of blue smoke it seems to leap from place to place on the leafy floor of the camp as the trees sway in the breeze high above. And when you listen you hear life in a myriad forms all around—birds, monkeys, bees; the rustling of a nearby stream and the never-ending voice of the forest itself. It continually whispers assurances that all is well, that the forest is looking after its children.

The heat of day in a village is unbearably, suffocatingly hot and oppressive, but in the forest it is just warm enough to make you want to stretch out and bask in the dancing, shady sunlight, like other living creatures. But the pygmies have learned from the animals around them to doze with one eye open, and a sleepy midday camp can become filled in a minute with shouts and yells and tearful protestations as some baby, crawling around this warm, friendly world, crawls into a bed of hot ashes, or over a column of army ants. In a moment he will be surrounded by angry adults and given a sound slapping, then carried unceremoniously back to the safety of a hut. It does not matter much which hut, because as far as the child is concerned all adults are his parents or grandparents, they are all equally likely to slap him for doing wrong, or fondle and feed him with delicacies if he is quiet and gives them no trouble. He knows his real mother and father, of course, and has a special affection for them and they for him, but from an early age he learns that he is the child of them all, for they are all children of the forest.

When a hunting party goes off there is always someone left in the camp. Usually some of the older men and women, some children, and perhaps one or two younger men and women. The children always have their own playground, called *bopi*, a few yards off from the main camp. At Apa Lelo it was where the Lelo twisted around an island and one branch cut in almost between Cephu's camp and the main camp. It was fairly shallow there, and all day long the children splashed and wallowed about to their heart's content. When tired of that they had a couple of vine swings in their *bopi*; one was a small one for younger children, and the other was hung from two tall trees. Infants watched with envy as the older children swung wildly about, climbing high up on the vine strands and performing all sorts of acrobatics.

There were always trees for the youngsters to climb, and this is one of the main sports even for those not yet old enough to walk properly. The great game is for half a dozen or more children to climb to the top of a young tree, bending it down until its head touches the ground. Then they all leap off at once, and if anyone is too slow he goes flying back upwards as the tree springs upright, to the jeers and laughter of his friends.

Like children everywhere, pygmy children love to imitate their adult idols, and this is the beginning of their schooling, for the adults will always encourage and help them. What else do they have to be taught, except to grow into good adults? So a fond father will make a tiny bow for his son, and arrows of soft wood and with blunt points. He may also give him a strip of a hunting net. The mother will delight herself and her daughter by weaving a miniature carrying basket, and soon boys and girls are 'playing house'. They solemnly collect the sticks and leaves, and while the girl is building a miniature house the boy prowls around with his bow and arrow. He will eventually find a stray plantain or an ear of corn which he will shoot at and proudly carry back. With equal solemnity it is cooked and eaten, and the two may even sleep the sleep of innocence in the hut they have made.

They will also play at hunting, the boys stretching out their little bits of net while the girls beat the ground with bunches of leaves and drive some poor, tired, old frog in towards the boys. If they can't find a frog they go and awaken one of their grandparents and ask him to play at being an antelope. He is then pursued all over the camp, twisting and dodging amongst the huts and the trees, until finally the young hunters trap their quarry in the net, and with shouts of delight pounce on him, beating him lovingly with their little fists. Then they roll over and over in a tangle with the net until they are exhausted.

For children life is one long frolic interspersed with a healthy sprinkle of spankings and slappings. Sometimes these seem unduly severe, but it is all part of their training. And one day they find that the games they have been playing are not games any longer, but the real thing, for they have become adults. Their hunting is now real hunting, their tree climbing is in earnest search of inaccessible honey, their acrobatics on the swings are repeated almost daily, in other forms, in the pursuit of elusive game, or in avoiding the malicious forest buffalo. It happens so gradually that they hardly notice the change at first, for even when they are proud and famous hunters their life is still full of fun and laughter. While the children are playing about in their *bopi*, safely out of the way yet close enough not to be able to fall into any real danger, the others who have stayed behind from the hunt have any number of ways of occupying themselves, if they do not want to sleep.

The first time I ever visited a hunting camp I heard its many sounds some fifteen minutes before I finally walked into the clearing. First I heard a sharp, hollow, tapping sound. Then the whole orchestra of children's voices, loudly gossiping women and chattering men blended with the hammering and occasional shouting and snatches of song. In the camp there was only a handful of people; but they were all busy, and when pygmies are busy they are noisy.

A youth was standing beside a fallen tree trunk, pounding away at a narrow strip of bark with the end of an elephant tusk, hafted simply into a wooden handle. He was making the bark cloth that is the main clothing of the pygmies, though more and more the men are getting used to the European-style clothes of the villagers. In the forest they still prefer bark cloth for several reasons, and it is easy to make and costs nothing. There are several vines that yield a suitable bark. After a strip has been cut it is laid over an outstretched leg while the artisan chops casually away at it, removing the outer bark. Then it is either soaked in water or mud, the mud giving it a bluish colour, or softened by smoking it over a fire. After that it is spread over a fallen tree trunk and beaten with an ivory tusk. It spreads out into a soft, supple cloth, the texture varying with the different types of bark.

There are many fashions in bark cloth, not only in the choice of a particular bark, but in the way it is dyed or decorated, and also in the way it is worn. The women nearly always do the decorating, using *nkula* bark to make a red paste, or the gardenia fruit, *kangay*, for black dye. And on certain barks the juice of a citrus fruit stains the material blue. As often as you see a man making cloth you will see a woman sitting down, legs straight out in front, as always. Her simple tools lie around her, and the bark cloth is spread over her thighs. If she is using the red *nkula*, first of all she has to rub two pieces of the wood together, with a little water to turn the powder into paste. Then, using a finger, she daubs the fresh cloth, which may be white, fawn, brown, or blue, with bright splotches of red. If she is using the fruit juice she carefully collects it in a cup made from the hard shell of some forest fruit, and with little twigs she draws fanciful patterns of straight lines criss-crossing the cloth with unconventional pygmy geometry. When the cloth is finished she throws it over the roof of her hut to dry, and looks around for something else to do.

Women usually work in twos or threes for company, and nearby a friend will be attending to some other womanly task. They spend hours at a time making the elaborate belts which hold the bark cloth in place. The material culture of the BaMbuti is minimal, and simplicity and utility are usually the main characteristics. But there is nothing simple about the belts they make. One belt takes several weeks in all. First the vine has to be collected, dried, and shredded. Then the strands are taken and tied around a stout twig stuck in the ground. The woman holds the stick firmly between her toes and pulls a number of the strands to form the core around which she braids an outer binding. There may be as many as two hundred braided lengths to a single belt, bound decoratively in the centre, each end worked into ties for fastening the belt tightly so that it will not only hold the bark cloth in place but also support knives, machettis, and any of the bundles of roots and other things the pygmies tie to their belts while on the trail. This leaves their hands and arms free.

The same vine, *nkusa*, is used for making hunting nets. After it has been shredded it is deftly rolled on the thigh, one strand being rolled into the end of the next so that when two strands are twined together a woman can present her husband with a length of stout cord long enough to add several feet to the family hunting net. This is virtually a full-time occupation, and one which both men and women indulge in whenever they have both the spare time and the inclination. There are few occupations that are strictly reserved for either men or women, and although a man will attend to his own net, repairing it and adding to it, a mother will often make a complete hunting net to present to her son when he gets married.

The other major activity that goes on almost every day in any pygmy camp is the repairing and alteration of the huts. New leaves have to be added, or the old ones changed to avoid leaks; and if relatives come to visit then either the house has to be enlarged or a new one built. Whenever this happens there are repercussions throughout the camp, since people will have to change the entrances to their huts to avoid looking directly into the stranger's hut. This also happens if there are any arguments between two families. The two disputants, though they may be next door to each other, will build little 'spite fences' between them, so that their huts are turned almost back to back, the walls in effect turning the doorways to face in opposite directions. This may make them face into someone else's hut, and if they are not in sympathy they in turn will turn *their* doorway, and so on. It is usually the woman who does this, as she builds the hut and it is considered as her property. In fact this is one of the strongest arguing points a woman has with her husband. I have seen a woman who has failed to get anywhere in a matrimonial disagreement simply turn around and start methodically pulling all the leaves off the hut. Usually the husband stops her half-way, but in this case he was particularly stubborn, and waited until she had taken all the leaves off, then remarked to the camp at large that his wife was going to be dreadfully cold that night. There was nothing for her to do, without losing face, but to continue; so reluctantly, and very slowly, she started to pull out the sticks that formed the framework of her home. By this time the camp was agog, because it had been a long time since anyone had seen a domestic argument carried quite this far.

The poor woman was in tears, for she was very much in love with her husband, and the final step, if he did not stop her, was for her to pack her few belongings and walk off, having completely demolished their home first. Then she would return to the home of her parents. It is the nature of a pygmy never to admit he is wrong, and the husband was beginning to feel equally anxious. Things had gone too far for either of them to patch up the quarrel without being shamed in the eyes of all who were watching, curiously, to see what would happen next. He sat silently, hugging his knees, looking as miserable as his wife. But he brightened up suddenly, turning around to see how far the demolition had gone. Only a few sticks had been pulled out, and he called to his wife and told her not to bother with the sticks, that it was only the leaves that were dirty. She looked

1 A group of men sets off for the hunt, some carrying their hunting nets, others spears or bows and arrows
2 Ageronga's wife prepares a bundle of *mongongo* leaves for thatching her hut

3 Masisi hunts birds and monkeys. The wooden point of his arrow is tipped with poison

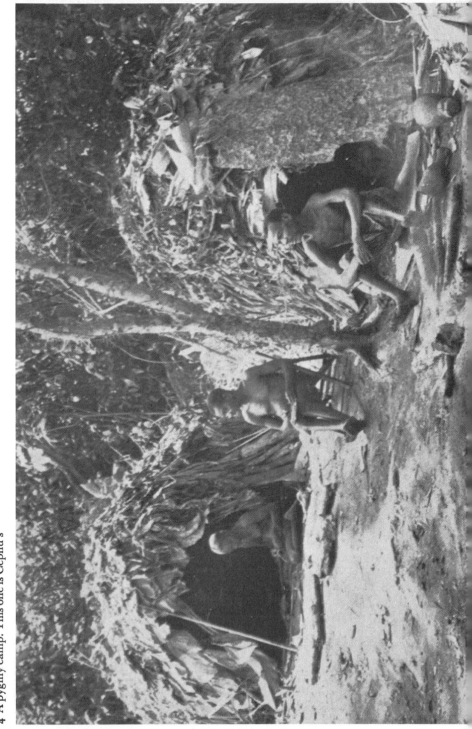

4 A pygmy camp. This one is Cephu's

5 A pygmy woman (right) helps a villager pound locally grown rice
6 above right Asofalinda's eldest daughter being painted
with *kangay* juice for a wedding
7 right Akidinimba (in hat) celebrates her forthcoming wedding
with the other girls of the *elima*

8 Tungana, a grandfather many times over, in gennet-skin cap

9 Kenge uses a hunting bow as a musical instrument

10 Old Moke (in a sloth-skin hat) listens, as Cephu tells a legend
11 Kelekome and his uncle Masalito examine a piece of honeycomb to see if the hive is ready for raiding. Honey is probably the most prized food in the forest

at him with a puzzled 'Ayiiiiii?' and then, understanding, asked him to help her carry the leaves down to the stream. This they did, and together they gravely washed every single leaf, brought them back, and while she put the leaves joyfully back on the hut he stoked up the fire and went off with his bow and arrows to see if he could find some game to bring back for a special dinner. The pretence was that she had been taking the leaves off not because she was angry, but because they were dirty and attracted the ants and spiders. Nobody believed this, but everyone was glad the quarrel was over, and for several days women talked politely about the insects in the leaves of their huts, taking one or two of them down to the stream to wash, as if this was a perfectly normal procedure. I have never seen it done before or since.

But apart from little excitements of this kind during the day, the camp really only comes to life when the hunt returns. Long before the first hunter appears the whole camp knows exactly who has caught what. They know partly by listening for the shouts and arm-claps of the returning hunters, which they can hear from a considerable distance. And they know partly because throughout the day children come and go between the camp and the hunt, if they are not too far apart, bringing news to and fro.

As soon as the hunters return they deposit the meat on the ground and the camp gathers to make sure the division is fair. Nobody acknowledges that it is, but in the end everyone is satisfied. Cooking operations start at once and within an hour everyone is eating. If the hunt has been a good one, and the day is still young, the most energetic men and women dance immediately afterwards, followed by the children. In the course of such a dance they imitate, with suitable exaggeration, the events of the day. Or if the hunt has not been so good or a man is tired and does not feel like dancing, he will sit down and gather his family around him and tell them something that has happened to him on the hunt, something wonderful and exciting, but which naturally happened while nobody else was looking.

Blessed with a lively imagination, these stories grow until it is difficult to tell the difference between what is meant to be a straight, factual account and a traditional legend. And there is always the element of the two-way process in which fact becomes legend, and legend becomes accepted, through constant repetition, as fact. The dream world of the pygmies is a very real one, full of hidden meaning and significance. There is one story that Cephu is still fond of telling, and he still tells it in almost exactly the same way as he told it six years ago when I heard him first tell it.

He had been into the village for a short visit, and we all knew he had gone empty handed. Yet when he returned a few days later he brought a sizeable load of plantains and rice with him. Everyone asked him who he had stolen it from, and Cephu, with a great smile spreading across his face, was silent and waited for his audience to settle down. With his wife and his children close beside him, and his friends all around, Cephu sat simply on the ground in his most charming and bewitching way, old rascal that he was, even then. He looked up at the trees and opened his arms and raised his expressive hands upwards. He began telling the following story in his beautiful, rich voice, the seriousness of what he was saying only betrayed by the smile that lurked in the depths of his shining brown eyes.

'I am only a poor MuMbuti, the forest is my father and my mother. But my negro master had asked me to bring him some meat.' So Cephu started, self-deprecating, using the term for his patron in a sense which only the negroes use it, as indicating ownership. The pygmies accept this attitude of superiority, outwardly, because it pays . . . as the story went on to show.

'My master asked me for meat, so I hunted and I hunted and I hunted, and I caught *sondu* and *lendu* and *mboloko* and *sindula*, and I cut them all up and dried them carefully. When they were nice and dry I wrapped the meat in fresh *mongongo* leaves and put it all in my basket and set off for the village. But on the way I saw a bad spirit coming towards me. I tried to escape but he caught up with me and demanded the meat. I said, "No, this is for my master, I can not give it to you. You may take all my own food, but not this meat . . . that is for my beloved master." But the spirit was not to be put off, and I had to fight with him. Finally I managed to escape, because I am strong and a great hunter, and I ran on towards the village as fast as I could. Spirits are very bad things to fight. They would kill any villager, only we pygmies can deal with them. But then I saw the spirit again and this time he was more clever than I, and he changed into my grandmother. "A'i, Cephu! Where are you going with all that meat?" she asked me. Now my grandmother died many years ago and I had to be polite to her. So I told her I was taking it to my beloved master. But she got angry and demanded it, asking if I would refuse it to her, a pygmy, my own grandmother, and bring it to a villager? I pleaded with her and begged her, telling her how good my master always was to me, and how hungry he was for meat. But she snatched the basket from me, and because she was my grandmother I could not strike her or stop her, and she ran off down the trail.

'As she was running she changed into a spirit again so that she could run faster, and as soon as my grandmother was a spirit once more I could fight, so I fought and fought, and the spirit nearly killed me, and he went off in the manner of spirits and left me with nothing, not even my basket.

'When I came to the village I was crying because I had nothing to offer my master, and when my master asked me what was wrong I told him this story. He was terribly sorry because he knows what dreadful lives we BaMbuti lead, surrounded by all these bad spirits. He said if it was not for the spirits he would come into the forest and hunt himself, but everyone knows how dangerous the spirits are to villagers. Then I asked him if I could have some rice as I had lost all my own food as well as his. He said, "Of course, my poor pygmy, you have tried to help your master, and you can't help it if you are just a pygmy and have to live amongst all those dreadful spirits."

'So he gave me rice and plantains and I came away happy. How kind the villagers are to their poor pygmies!'

Every time Cephu tells this story it brings knowing smiles and laughter to the faces of all who listen, because every 'poor pygmy' has some kind of trick like Cephu's by which he twists his 'master' around his little finger, and at the same time heightens the fear of the villagers for the forest by constantly repeating stories of imaginary spirits and dread dangers.

This story, as often happened, set off another. Old Asofalinda, Ekianga's sister, was not to be outdone. She described
how she had cut down a fine stalk of plantains outside the
village, but was surprised in the act by the owner. She just
had time to push the plantains into a hole in the ground at the
foot of the tree. The villager asked what she was doing there,
and what she was putting in the hole. 'Oh, master!' she replied.
'There was a dreadful, dangerous spirit-animal and he was
going to attack you so I drove him into the hole and I dare not
leave it or he will come out and kill you! Run and get your
spear and save me, your poor pygmy!' The villager was scared
out of his wits and ran away. Asofalinda promptly hid the
plantains elsewhere and quickly returned to the hole.

At this point in the story, Masamba, Njobo's wife, came to
help act out the rest. Asofalinda crouched at the foot of a tree,
pretending that she was covering a hole in the ground, and
Masamba crept on to the scene carrying a spear, and looking
very frightened, acting the part of the villager. The way it was
acted the villager came up behind Asofalinda and asked timidly
if the spirit-animal was still there. 'Yes,' said Asofalinda, over
her shoulder. 'But I have used my magic and sung it to sleep
so that you will be able to kill it. But you must be very quiet
and very careful. You are the only one who could do it, you
are so big and strong and clever, not like we poor pygmies.'

Saying this she slowly uncovered the hole and whispered to
the villager to stoop down and peer in. Masamba acted her
part, and cautiously bent over and peered into the imaginary
hole. 'No,' said Asofalinda. 'You will have to look deeper, it is a
big hole and the spirit-animal is right at the back. So Masamba
stooped lower and Asofalinda acted the rest of the scene with
no further commentary. It was hardly necessary. She crept
behind Masamba and kicked her heartily in the backside, and
ran off clapping her hands with mirth and laughing. Even
Masamba, who suffered the brunt of the joke, enjoyed it so
much that the two of them re-enacted it four times that one
afternoon.

In ways like this, even on a rainy day when there is no hunting or gathering to be done, a camp can be made lively and
entertaining. But generally rain brings depression. Everyone

sits in their hut and peers out miserably. On the whole the
tenor of life is even, wet or dry, and with few real ups and
downs that affect the life of the group as a whole. Occasionally
an elephant is killed, and that means the entire camp will be
abandoned at once and a new one built around the dead animal, so that it can be eaten on the spot. The pygmies point out
that this is easier than bringing the elephant to the camp. It
means a week of merrymaking, of singing and dancing and no
hunting. The only trouble is that it also means visits from the
villagers. On such occasions they will pluck up their courage
and venture into the forest, feeling that with such an abundance
of meat the pygmies will be in a generous frame of mind. The
pygmies try to counter this by immediately sending meat to the
village, as they dislike intruders in their forest. But when intruders do come, they are generally courteously received and
allowed to stay for a day or two. After that, if they are slow with
gifts of rice or tobacco, the pygmies have various ways of making
them move on. The forest is the world of the pygmy, and he is
not bashful in telling a villager who outstays his welcome just
what he thinks of him.

I have seen villagers pass through a camp we had that was
only a few hours from the village on the way to a negro fishing
ground. We were constantly beset by villagers passing through,
and after a while the pygmies got tired of feeding them, tired
of being used as a convenient stopping place. So when they
were asked for food the pygmies would agree on the condition
that the villagers joined them in a game of *panda*. This is a
gambling game where beans are thrown on a mat, or on a flat
piece of ground, and some of them are quickly scooped up by
one of the players. The other, looking to see how many are left,
at a glance, has to estimate how many his opponent has taken.
He then tells his opponent to throw down one, two, or three
beans, or to stay as he is. The count is then made—also at a
glance—and if the player has an exact multiple of four he wins.
But the pygmies always win. For one thing they have an un-
canny sense of form that enables them in a moment to know
the exact number thrown down. And for another, the villagers
are not as nimble with their fingers and toes as the pygmies.
Because when a pygmy sees that he is going to lose by a couple

of beans, he quickly drops them from the store he has concealed between his fingers, or even in his hair, so that with a triumphant toss of his hand or his head he can throw in the winning bean.

The villagers left that camp stripped of their fish, their tobacco, and whatever else they had to gamble away; and they soon came to prefer an alternative route by-passing our camp. There are a few villagers that are welcome, but they are rare. Others can come with sackloads of food and still get a cold welcome. I have seen this happen, and as soon as the pygmies have divided up the food they send the villagers on their way. But others can come empty handed and be received gladly, because they understand the pygmies and the forest. For the pygmies this is the criterion of a 'real person', as opposed to the animal-humans who live in permanent villages out in the open, away from the shelter and affection of the trees. But any such visitations are temporary, and they usually have some political significance. A negro chief who wants to win more pygmies to his village may send out a son with gifts to stay in a pygmy camp. Every day more gifts and food are sent out so that he can distribute these. It is a risky business because, as the villagers say, 'The BaMbuti eat us up until we have nothing left for ourselves!' But if a son wins the affection of the pygmies, his father will have won a steady supply of meat—for a time anyway.

One afternoon I saw two village girls arrive, followed by two servants who threw down loads of plantains and sweet bananas, and a sack of manioc flour. The servants left and the girls came over quite openly and sat down unconcernedly at Masamba's fire. Masamba came out of her hut and greeted them with warmth, and soon there were a number of women gathered around, chattering away gaily and asking for news of the village.

The girls were Amina, daughter of an important BaBira sub-chief, and her cousin. Her father was hoping, she said, that she might be allowed to stay for a few days and bring him back some meat. She obviously knew the pygmies well, because without waiting to be asked she distributed the plantains and said she would keep the manioc flour until the meat began to come in.

Apparently she was well liked, because as soon as people heard of her presence they came over to shake her hand, in the manner of villagers, grasping each others wrists. Kenge told me all about her, and how her father was one of the least animal-like villagers. He asked me which of the two girls I thought looked the more attractive. I said Amina, without a doubt, because she was the most beautiful girl I had seen in a long time, and the BaBira are not noted for their good looks by any means. Amina was tall and a warm brown in colour. She carried herself with the inborn grace of Bantu women, her eyes were deep and thoughtful, her expression was kind and gentle. Yes, by all means she was the more beautiful of the two. Kenge nodded thoughtfully and changed the conversation.

That night I sat up late with the men, and when I went back to my hut I was surprised not to see any bachelors sleeping on the floor as they usually did. But three logs were placed beside the bed, and were glowing cheerfully. I sat down on the edge of the bed and found, as I had expected, that Kenge had decided it was too cold to sleep on the ground all by himself, so he had curled up in my blanket. He was considerate enough, on such occasions, not to take my blanket on to the floor, but rather to take half of the bed. So I crawled in beside him and was about to say something when I realized that the body beside me was considerably taller than Kenge, and very differently shaped. It was Amina.

It was a difficult moment, but I could see the whole situation at once . . . I had become the centre of a nice political manœuvre. The Chief had sent his daughter, not to win the pygmies by her gifts of plantains and manioc flour, but to win them by winning me. The pygmies, anxious to get what they could out of the situation, liking Amina anyway, but having some consideration for me, got Kenge to find out which of the two girls I preferred. That, evidently, was to be the limit of my choice in the matter. I confess I was glad I had chosen Amina, but I could see all sorts of difficulties arising if I were to take advantage of the situation. No doubt the Chief had in mind the considerable bride-wealth he could demand should, by any chance, his daughter bear a mulatto child. Still, it was good of him to send his prettiest daughter.

I considered for a minute or two and then said, 'Amina?' very quietly. It had begun to rain, and I half hoped she would not hear me. It would be a terrible disgrace to her—and to me —if I turned her out, yet I wanted no part of a lengthy and costly dispute over bride-wealth. But Amina was by no means asleep. She laughed lightly and made fun of my KiNgwana name, which means 'the long one'. So I said, 'Amina!' once again, a little more sharply, still trying to think what a supposed gentleman and scholar should do under such circumstances. But Amina was not helping. She just snuggled closer and said, 'Yes, tall one?' Then in a moment of inspiration I simply said, 'Amina, the roof is leaking.' And with masculine authority I added, 'Get up and fix it.'

Without a word she got up and started searching for the leak, pulling at the leaves until a steady trickle of water came down, splashing on the floor beside the bed. By the time she had repaired the damage and checked the whole roof with womanly efficiency I was asleep—at least so I pretended. With a sigh she pulled the blanket from me and also went to sleep—or perhaps she, too, was pretending. But it was a solution, if a chilly one, and my last thoughts that night were that Kenge would at least have left me some of the blanket, no matter how annoyed with me he was. These villagers—they were just animals.

When I woke in the morning I hardly dared open my eyes; but a tentative exploration proved that the other half of the bed was empty. I thought that all was well after all. Amina had cut her losses and left at dawn. But when I had put my trousers on and stooped through the low entrance of the hut into the camp, there she was. She was sitting on a log beside the hut, quietly fanning the fire and cooking my breakfast, just as every other wife was doing for her husband. Kenge was making it very obvious to everyone that he was eating with the bachelors, and they were making it obvious that they had not slept in my hut that night. Everyone else studiously ignored my appearance except old Moke, who was walking past on his way to the stream to bathe. He looked at me sideways, shook his head and gave a little knowing old laugh. And as he walked on he shook his head all the more vigorously and laughed all the louder.

I was about to go after him, thinking that at least I could

ask his advice, but Amina caught hold of my trouser leg and said, 'There is no need for you to go to the stream to bathe, I have brought you water—behind the hut.' So I went behind the hut and washed, and then sat down and had breakfast while Amina went to join the other women and chat to them. While she was gone Kenge strolled by and elaborately asked me if the food was all right. He also asked me, rather loudly, if I had slept well, and Amina answered from across the camp, 'Yes, he told me to mend the leak in the roof!' This proof of domestic bliss brought howls of mirth from all around, the more infuriating because I could not see what was funny about it. Then Kenge told me that this was what men always said to their wives when they wanted to be allowed to sleep in peace.

That night, Amina and I came to an agreement. She was to stay with me and cook my food . . . and mend leaks in the roof every night. This way she preserved her reputation and I mine, and there would be no complications. And so it was. I grew very fond of Amina, and later on, when I was in a village and became very ill, she walked ten miles every day to see me, and ten miles back. And she brought me refreshing fruits and sweet-smelling plants and flowers, and sat beside my bed without saying a word. Only once, I woke up and found she was sitting on the edge of the bed and holding my hand in hers. But when the pygmies moved camp, Amina had taken the opportunity to go back to her father's village. I was never sure whether I was glad, or not more than just a little sorry to see her go.

But when I thought about it, apart from all the other reasons, I realized that Amina could never have stayed. For no matter how well she was liked by the pygmies, she was still a villager. She had not been brought up to do the hundred and one little things that a pygmy woman has to know how to do. She did not even know how to beat on the hunt, let alone which vines to follow to find the delicacies hidden in the ground by their roots. She belonged to another world.

Chapter Eight

THE RIVER LELO flowed on, rising and falling endlessly; the men went out to hunt and returned with meat; the women went out to gather and came back with mushrooms and nuts and fruits, and the children laughed and played. Old men and women did a little foraging in the vicinity of the camp, then sat in the cool shade and told each other about all the things they had done when they were young and would do all over again if they had the chance. At Apa Lelo one day followed the last in this happy-go-lucky way as though this was all there was to life.

But at the central fireplace there was always a basket of food, showing that it was a *kumamolimo*, the only outward indication that the God of the Forest had been invoked and was waiting silently nearby, in the trees or in the Lelo itself, ready to enter the camp at night and join his people in their songs of joy and praise. The casual observer might have thought the youngsters were playing a game when they went around from hut to hut with a fishing line and a basket, collecting food from everyone. There was certainly nothing about their demeanour to indicate that this was an essential part of a great act of communion. And if the same observer were told that this was so, he would probably form the opinion that the *molimo*, whatever it was, could not mean very much if it was treated so lightly and casually.

This, in fact, is the opinion of the villagers. For them the ritual act is the important thing. In their religious rites it is the correct performance of the ceremonial that counts most. If correctly performed it is bound to bring the desired results, regardless of the accompanying thought, if, indeed, there is any. All attention is focused on the act itself. And it must not only be performed correctly, but with a solemnity that betrays fear.

The difference between the two attitudes is perhaps one of those fine shades of difference that divide magic from religion, though there is so much overlapping that they can seldom be divided in practice. Whereas the villagers believe that the act itself brings about results in a way they cannot explain, which is what we call magic, the BaMbuti do not believe in this at all. They believe in a benevolent deity or supernatural power which they identify with the forest. To this they owe as much respect and affection and consideration as they owe to their own parents, and from it they can expect the same in return. So, for the pygmies, it is not so much the act itself that counts, but the manner in which the act is performed and the thought that goes with it. The collecting of food from hut to hut is not a magical act, but a way of emphasizing that this is something in which every single member of the group must participate.

At first I had thought that the *molimo* was virtually the sole concern of the men. In the evenings, when the singing first started, there were sometimes a few girls who stood nearby and danced amongst themselves as we sang; but they were soon sent to join the rest of the women and children in their huts . . . certainly long before the *molimo* made its appearance. But one day I felt that something different was going on. I had noticed that Kondabate, Manyalibo's daughter, who was married to Ausu, was building an extension at the back of her hut. She and Ausu had no children, and when I asked her why she was enlarging her home she said there was going to be an *elima*, and that all the girls would stay in her hut, under her care. An *elima* is a celebration that takes place when a girl reaches maturity, marked by the first appearance of menstrual blood. At such a time she enters a special hut with her girl friends and stays there for a month. But I had heard none of the usual discussion about the event, and I felt this could not be the full answer. Then, that afternoon, an old woman arrived with her husband. They stopped first in Cephu's camp, where they had relatives, but then came right on into the main camp, where they were received with the greatest respect. The husband sat down at the *kumamolimo* and was offered food; the old woman disappeared into Kondabate's house. She stayed there until evening, but all I could hear from inside were occasional snatches of song. It was plain that there were several girls in there as well, and it sounded as though they were either being taught new songs or else were rehearsing songs they already knew. They never sang more than a few bars at a time, and I was continually being startled, thinking that I heard something resembling one of the *molimo* songs.

That evening the old man joined us for the *molimo* singing. He was given special recognition by Moke, who offered him his chair and then sat on the ground beside him. Old Tungana seemed to know him well too, and the elders, the *Mangese*, the Great Ones, formed a tight cluster of their own, apart from the hunters. The hunters formed their own group, and the youths theirs. The youths and a few of the hunters danced, and we were surrounded by women and girls who watched and joined the songs with handclaps. The girls formed their own dance circles, but they seemed to be dancing with much more intent than on other nights, and they were allowed to stay up until much later. Then they were sent to bed, and the singing began in earnest. All the girls that had danced went into Kondabate's hut, and as they opened the leaf door to enter I saw that the old woman was sitting inside, hunched up over the fire, her eyes as bright as the coals, wide and unblinking.

The singing went on particularly late that night, and it was only an hour after Moke's usual statement that he was tired that we were awakened by the blasting of both *molimo* trumpets as they entered the camp. I got up hurriedly, and saw that Kenge was already gone. At the door of my hut someone ran up in the dusk and pushed me back, telling me not to come out, it was too dangerous. Just as he said this there was a tremendous blow on top of the hut, and the *molimo* swept by. It was surrounded by at least twenty youths, all armed, as far as I could see, with long branches with which they beat on the roofs and walls of every hut. They continued this for an hour, the trumpets blaring angrily, the youths making loud hunting cries. As the forest grew lighter I could see that all the youths were decorated with leafy branches stuck in their belts, so that it looked as though each trumpet was surrounded by a little forest of its own.

Quite a number of huts had been damaged, but there was no word of protest. Ekianga danced up and down with a spear under each arm, and ran rings around the trumpets. I was more impressed than ever by the high-stepping gait that he and everyone else used. They did not simply run, they seemed to float over the ground at tremendous speed, their feet high in the air. I saw why I had been told not to come out, because the youths that were armed with sticks and branches on the flanks of the trumpets kept on lashing out viciously, at imaginary intruders. When Ekianga came too close, or any of the others who were running about independently, beating on roof-tops and tearing off leaves, they had to leap from side to side to avoid being hit. Each person seemed concerned only with what he was doing. Nobody so much as looked at anyone else.

Then the rest of the men came out of their huts, one by one, and went over to the *kumamolimo* and began singing. It was no tentative, sleepy sing-song, as it was on many mornings, but singing in real earnest, as loud and as vigorous as at the height of the night. The *molimo* fire was fanned up into a blaze, and youths ran swiftly around the camp picking up bits of wood from everywhere, dancing back to the *kumamolimo* and throwing them on to the fire. Then, with a wild rush, the two trumpets joined together and were surrounded by a single dancing forest, and blowing a martial air they tramped around the edge of the fire and disappeared into the trees. They seemed to cross the river and climb up the hill on the far side, until the sound of our own voices drowned the more distant music of the trumpets.

For the whole of that day everyone's nerves seemed to be on edge. Asofalinda was in a rage about something and she strode up and down the length of the camp, shaking her arm and pointing at everyone in turn, telling them they were all lazy good-for-nothings. Masisi was in one of his worst tempers and lashed out with an acid tongue at anyone foolish enough to try and talk to him. Manyalibo sulked and said that it was pretty poor that it should be so difficult to fill one *molimo* basket. There should be two, three, or even four baskets, and they should be full all the time, he said. He made some remark about Masamba not having done much to help, so Masamba threw an empty basket at him and said, 'There, you make so much noise, go and fill it yourself.' Even before the hunt went off there were two full baskets hanging up, and by evening there were four.

The change between the atmosphere of the morning and the

evening was unbelievable. Usually when a day starts off in such ill humour it continues in that way to the bitter end. But the hunt seemed to restore everyone's good spirits, even though it was not outstandingly successful. I did not go on the hunt, but stayed in camp. And all day long the girls were singing in Kondabate's house, and the old woman was with them.

After we had eaten the evening meal Manyalibo went around the camp himself and took a piece of wood from the fire outside each hut, placing them all on the *molimo* fire. Then, before it was completely dark, the men gathered to sing. Even Cephu was there, and the singing was light and cheerful. The girls that had been in Kondabate's hut came out as soon as we started and I could see they had painted their bodies with black *kangay* juice, and in their hair they had twined circlets of vine, one strand of which stuck way out in front, over their foreheads, a small bunch of feathers dangling from the end. They kept together in a tight little group, their bodies swaying in unison to the songs of the men. Then they began to dance, still so close together that I could not see between them. They danced all around the group, forcing us closer and closer to the fire. In a snaking, graceful line they seemed to take charge of the music, increasing the tempo until the men were as exhausted as they were, and we all stopped for a rest.

By then it was dark, and I expected the girls and the rest of the women and children to retire to their huts. But they had built a fire of their own at the edge of the *molimo* group, and there they all sat, chatting inconsequentially as though they did this every night. It was then that I first noticed the old woman. Under a great headdress of vine and feathers her sharp, hatchet face looked more angular than ever, thin and straight, with eyes still wide and unblinking and, it seemed, unseeing. She sat in the middle of the group, hugging her knees. Her skin sagged in withered folds, her bones stuck out in ugly lines and bumps. She made a strange contrast with the background of young feminine beauty and vitality. The only time she as much as moved was when Kondabate joined the group.

Kondabate, the belle with filed teeth and great diamond-shaped scarifications on her stomach; Kondabate, the beauty who after two years of married life still preserved her fine, upstanding breasts and trim figure. But Kondabate, the most beautiful wife in the forest, was also Kondabate the childless. She too had on a headdress like the old woman, and her eyes, too, seemed unnaturally wide and staring. She was smiling and happy—until she sat down opposite the old lady. That was when the old woman moved, looking up from the fire into Kondabate's eyes. In an instant Kondabate seemed to lose her smile and her happiness. She became as empty as the withered old shell of a woman in front of her, and she sat as quietly, staring into the flickering flames.

After a while the men started singing again, but gently, and then with a shock I realized that the women were singing as well, the sacred songs of the *molimo*. And they were not just joining in, they were leading the singing. Songs that I had thought only the men knew and were allowed to sing; all of a sudden the women were showing that they not only knew them but could sing them with just as much intensity. Kondabate suddenly sprang up and seized the *banja* sticks from Kelemoke, and began beating the urgent *molimo* rhythm: *one* two, *one* two three, *one* two three. She beat the *banja* faster and faster, and I saw the old woman begin to twitch at every muscle. Kondabate was already half dancing, and it was as though she was passing the life and force from her body into the dried-up, skinny old body in front of her. The old crone's legs started stamping up and down, and she sat back on her buttocks so that she could lift her feet higher. Then her arms began moving, pointing in this direction and that; even her fingers each seemed to have a dance of its own. But her head was still, and her eyes were fixed on those of Kondabate.

At this point Kondabate handed the *banja* to another of the women standing nearby, and she led the young girls right into the centre of the men's group, dancing in a close circle around the *molimo* fire. As they moved around the men stood up and made room for them, and now the singing came almost entirely from the women. The men just stood and stamped and clapped and growled a bass accompaniment. The girls danced more and more wildly, but their heads and eyes were always turned towards the *molimo* fire, no matter how violent the body movement. After nearly half an hour the old woman, still crouching, ran into the centre after the girls, and turning to face them from the far side of the fire she drove them slowly backwards, back until they were lost amongst the others at the edge of the group. The singing was quieter, and in a few minutes it faded away altogether. Not a word was spoken.

After a short rest the singing started again, and this time there was no mistaking it. Manyalibo's wife picked up the *banja* and beat them, sharply and decisively, and the girls began the song. The men followed in obedient chorus. Once again, as the song grew faster, the old woman began to twitch until her whole body was alive with movement. Then she sprang up and came into the centre of the *kumamolimo* alone. For an instant she stood there, and her head moved sharply from side to side. It was like the movement of a bird, keeping a wary eye open for possible foes. As she began to dance she seemed to sink slowly into the ground, her knees folding up like a concertina, her body bending at the waist; but her head always stuck out in front, jerking from side to side, eyes staring and expressionless.

She circled the fire three times, looking out and upwards into the night, but unseeing; then, as if satisfied that she was alone and unobserved she flung her head and her whole body around and faced the fire directly. She was about seven or eight feet away from it. She advanced slowly and cautiously, two steps forward, one step back, her attention riveted on the flames. The singing was louder and faster, but she refused to be hurried. In her own time she approached closer and closer until it seemed impossible that she could stand the heat. She hesitated at the brink of the fire, her frail body gleaming with sweat, her face glowing and trembling. She made as if she was going to pounce on the fire, but whirled around at the last moment and danced quickly back to where the women were singing, watching her, eagerly and anxiously. She sat down on her haunches and accepted a steaming bowl of *liko*.

The singing did not stop this time, but went on quietly at a slower tempo. The men made no attempt to move back closer to the fire. After about ten minutes the old lady, apparently refreshed and as full of vigour as ever, danced back into the men's circle, with Kondabate at her side. Their movements were perfectly co-ordinated, even to the twitching of their heads; as they sank down close to the ground they might have been one person. But as they neared the fire, once again having circled it several times, the old woman sprang to her feet and ran swiftly around to the far side and crouched there, staring at Kondabate through the flames. They started circling once more, each keeping her eyes on the other, moving from side to side so that they were always opposite each other. But now they were closer to the fire than they had ever been and on their hands and knees they passed through the hot ashes, as though trying to reach each other through the flames.

Finally Kondabate retreated, and squatted down at the edge of the circle, her body twitching just as the old woman's had done, her eyes staring and empty as she gazed at the *molimo* fire. The men heaped more wood on to the hearth, which was beginning to die down, and now they were singing as loudly as the women, stamping on the ground with their feet and dancing with their arms and bodies, but never moving from where they stood. The old woman grew taller and more upright. Her bent old frame straightened out and she stood proudly, arms at her side, bent at the elbows so that her hands were out in front. Those hands had a life of their own, pointing and gesticulating and dancing with infinite grace at the end of motionless arms. The woman no longer looked from side to side. She knew she had nothing to fear now, that she was the victor. And when she stooped once more to approach the fire she did not stoop so low, and each step was a step forward. As the singing quickened so did her steps, until with a burst of frenzied shouting she was driven right into the flames.

She seemed to hover there an instant, that skinny old crone who should have been burned to a cinder in a flash. But she whirled around and kicked out with her feet, scattering the sacred *molimo* fire in all directions. Blazing logs and glowing embers alike she scattered, right amongst the circle of men surrounding her. And then she danced away, erect and proud. The men, without even faltering in their song, quickly gathered all the scattered embers and threw them back on to the remaining coals, and then for the first time they moved in a dance of their own. They danced in a wild circle, and as they danced their bodies swayed backwards and forwards, facing the fire, as though by imitation of the act of generation they were giving the fire new life. And as they danced the flames slowly began to rise, and as they rose the men danced all the more violently until they had brought the fire of the *molimo* back to life.

Twice more this happened, and each time the old woman made a more determined effort than before to stamp the fire out of existence. And each time the strangely beautiful and exciting, erotic dance of the men gave it new life. Finally the old woman conceded defeat and retreated amongst the others. Shortly afterwards all the women disappeared, and the men rearranged themselves at the *kumamolimo* without making any mention of what had happened. I looked at Moke, and found him staring at me curiously. He understood my question and nodded. Yes, this was what he had told me I was yet to see. There is an old legend that once it was the women who 'owned' the *molimo*, but the men stole it from them and ever since the women have been forbidden to see it. Perhaps this was a way of reminding the men of the origin of their *molimo*. There is another old legend which tells that it was a woman who stole fire from the chimpanzees or, in another version, from the great forest spirit. Perhaps the dance had been in imitation of this. I did not understand it by any means, but somehow it seemed to make sense. The woman is not discriminated against in BaMbuti society as she is in some other African societies. She has a full and important role to play, and there is relatively little specialization according to sex. Even the hunt is a joint effort. A man is not ashamed to pick mushrooms and nuts if he finds them, or to wash and clean a baby. A woman is free to take part in the discussions of men, if she has something relevant to say. In fact it was only the apparent absolute dominance of the male in the *molimo* that had been the exception.

But now even in this the woman had come into her own. She had asserted her prior claim to the fire of life, and her ability to destroy it, to extinguish life. Or had she been destroying it? Perhaps when she was kicking it out in all directions amongst the menfolk she was giving it to them, for them to gather and rebuild and revitalize with the dance of life.

I was trying to puzzle it all out when the singing started again, and I saw that the old woman had returned, alone. All the others were in their huts, and the men seemed to be singing as though by their song they could drive the last remaining female away. They sang louder and harder, but she kept circling the *kumamolimo*, keeping in the shadows, until with surprisingly swift and agile strides she was once more in our midst. In her hands she held a long roll of *nkusa* twine, the twine used for making hunting nets. The men continued singing, and as they sang the old woman went around knotting a loop around each man's neck, so that in the end we were all tied together. The men made no attempt to resist, rather they ignored what was going on. But when they were all tied up they stopped singing.

Moke spoke . . . I am not sure whether it was for my benefit or because it was his place to say what he did. He said, 'This woman has tied us up. She has bound the men, bound the hunt, and bound the *molimo*. We can do nothing.' Then Manyalibo said that we had to admit we have been bound, and that we should give the women something as a token of our defeat; then she would let us go. A certain amount of food and cigarettes was agreed upon, and the old woman solemnly went among us again, untying each man. Nobody attempted to release himself, but as each man was untied he began to sing once more . . . the *molimo* was free. The old woman received her gifts and went back to Cephu's camp, where she and her husband were staying. The couple stayed for another week or two, but she only danced once again. Before she left us she went to every man, giving him her hand to touch as though it were some kind of blessing.

It was not long after this that everyone started talking about when they should bring the *molimo* to an end. There seemed to be no specific length of time that it should run, and it was just a question of convenience. There was the matter of the *elima* to be considered too, as evidently there really was one that was due to be held for two girls, each of whom had reached her first period. The main issue at stake was not the fulfilling of ritual necessities, but rather how to make the best out of the festive occasion. There were different schools of thought as to whether it should take place in the village or the forest; the younger men held for the village, the older for the forest. All the older men except Tungana, that is, and the wily old grandfather said he thought the village was better as they might be able to make money out of tourists who would pay to see them dance. Tourists were increasing in number ever since the animal station had been established at Epulu, and the pygmies found tourists even easier to trick than villagers.

It was eventually decided that the *elima*, which had already begun in Kondabate's hut, should go to the village and take over Ekianga's house, as it was the biggest and he was related to one of the girls, anyway. The villagers would give food to the *elima*. Asofalinda would be in charge, everyone else would stay a while longer in the forest. Then they would bring the *molimo* to the village one night and have a final dance there, because the villagers believed that a period of mourning, which is how they regarded the *molimo*, should end with a ritual feast . . . and so they would give the pygmies more food. That would put all the men in a good humour to attend to their part of the *elima* festivities. I was a little surprised that they should even think of bringing the *molimo* to the village, but Moke told me not to worry, that no villagers would know anything of what went on in the forest, that if any villagers came to watch they would only see the pygmies imitating what the villagers do themselves. But he shook his head and added, 'But even so, it *is* bad. It is a thing of the forest, and it should end there. Perhaps we shall have to end it twice.'

The night before we were due to move back to the village everyone was very quiet. The women prepared the food for the evening meal without talking to each other. Tungana sat outside his hut, hunched forward, his grey old head resting on one hand because he had an abscess on the side of his neck. Moke wandered restlessly from one end of the camp to the other, and Masisi snapped at everyone who gave him an excuse. This was some three months after the *molimo* had begun, and we were in a camp much closer to the village than we had been at Apa Lelo . . . only about two hours' walk away. Cephu, as usual, had his camp at a distance, but rather further removed than usual, and we saw less of him than ever. The next morning we found that he had abandoned his camp the night before and gone to join a hunting group further to the west; he wanted no part of returning to the village.

Even the children were depressed, and sat around in silent little groups fitfully playing with spinning tops made from slices of forest nuts. Every now and again one of them got in the way of an adult and was severely smacked. His cries brought only a sterner cuffing, and by nightfall the camp was filled with wailing children and bad-tempered adults. Only the youths seemed cheerful. For them the village meant tobacco, palm wine, the wild, erotic dancing the pygmies love to perform for the villagers, and a relaxation from the tensions of the forest. They would no longer be hunting so they would be free from the restrictions that way of life imposed. They would no longer be in the forest, so they would be free from its ever-present authority. If the villagers thought the BaMbuti then came under their authority, that was their mistake.

While the youths dreamed of all the fun they were going to have, particularly with the girls in the *elima* house, all the adults were silent. Kenge, who although he was older, still liked to associate with the youths, stood boastfully in the middle of a group of admirers and talked of all the village girls he was going to sleep with. And not only one a night, but two, three, perhaps even thirty. 'And I am going to drink palm wine until there is no more to drink. I am going to steal all the palm wine there is in the village and drink it all myself, to make me strong. But I won't get drunk, so when they come to me and say, "Kenge, why have you drunken all our palm wine?" I shall just say, "You committer of incest."—Kenge had a very descriptive word for this—"If I had drunk all your palm wine would I be able to stand?" And I shall give them money to go and buy themselves beer.'

'And where will you get the money?' asked young Kaoya, trying hard to believe that his idol was telling the truth.

'Their wives will pay me for sleeping with them, of course!' answered Kenge, and he went into a dubious pantomime illustrating just how he was going to do this.

Manyalibo, who had been more morose than anyone, and was lying on his hunting net at the *kumamolimo*, propped himself up and shouted across the camp. 'You, Teleabo Kenge, you speak of the great man you are going to be when you get to the village. How is that you cannot be such a man here in the forest? Why do you spend all your time with children? Why do you leave the *molimo* at night and go to sleep like a suckling baby? How long have you been out of your mother's stomach? Come here at once and beat the *banja* for the *molimo*.'

This was one of the few occasions when Kenge did not answer back. He hesitated a second, then called out, 'I am coming, Father,' and came running over. He picked up the *banja* and started beating them for all he was worth. He called out to all the men to listen to him, not to be so lazy, to come to the *kumamolimo* and sing. Was not he, Teleabo Kenge, beating the *banja*? And was this not the last night of the *molimo* in the forest?

The singing that night was perhaps the best it had ever been. The *molimo* sang the most wistful songs, and the men answered it with love. When some of the youngsters began to dance Moke shook his head and told them to leave that for the village. Tonight was for singing. And just before dawn, when the *molimo* was about to leave the fireside, it sang to each of us in turn. It caressed the melody with love, and passed over our heads, one by one. As it left, Moke stood up and came around the group and paused in front of each of us, his hand fluttering lightly around our bodies, occasionally touching us as gently as the breeze. When he had finished he stood by the fire and sang to the forest by himself, with the *molimo* answering from a great distance, getting fainter and fainter. Finally they were both singing so quietly that I could hardly distinguish their music from the music of the crickets and the frogs and the birds of the night. Somewhere a hyrax gave its throaty, rasping call of despair; and an owl hooted. Moke sat down, and nobody spoke, and nobody sang. Something wonderful had come into our lives and filled them with the magic of love and trust, and though we still possessed that love and trust, we regretted the going of the power that had brought it.

At dawn we were awakened as usual by the sound of the two trumpets. But there was a quality about their tone that was harsh and brash. We all knew that the real *molimo* had gone, but we had to go through with the final ritual for the benefit of the villagers. The bellowings did not last long, and soon they disappeared in the direction of the village with a bunch of noisy youths in attendance. Later, when the main body of the camp was packed up and ready to leave, one of the youths came running back and told us with great excitement what fun they had had frightening the villagers with the *molimo*, but now it was safely hidden in the forest, on the outskirts of the village.

We got a warm reception as we emerged from the plantations amongst the neat rows of mud houses. The villagers eyed the baskets on the women's backs and tried to estimate how much meat they were bringing. They called out offers of free meals in an attempt to bribe their way into the favour of people whom they otherwise regarded as slaves. As we filed from the leafy plantation, leaving the forest behind us, we left a kind and friendly world behind us and entered another. Here, if the world was open to the sky it was also open to greed and suspicion and treachery. Village women attached themselves to 'their' pygmies and tried to pry into the baskets to verify just how much meat there was, accusing them of having eaten it all themselves. But the pygmies paid no attention. Some made for their own village and dumped their belongings there, but I wandered on with the others who went straight to their patrons to whom they presented the meat they had brought with them. Before long they were sitting down in whatever shade they could find, fanning themselves to keep the flies and mosquitoes away, eating heartily from enamel dishes, drinking dirty village water from enamel mugs.

After spending some time, rather like the pygmies, wandering around and greeting the various villagers, I returned to the pygmy village and sat in the shade of my veranda looking down the dismal, double row of tumbledown huts. About half a mile away, in the background, the forest raised its tall, dark face against the edge of the plantations, dwarfing the leafy plantain trees that were only three times as tall as men, making them look like blades of grass. Ineffectual blades of grass that had, with the help of village-man, usurped the place of the great primeval forest, but which in a year or two would be choked to a final death as the forest claimed its own. Clouds of dust arose as a small, disconsolate group of children played football with a ball presented to them by some fond villager. The dust got into my nostrils and made me feel bad tempered. Kenge,

THE FOREST PEOPLE

who was trying to cook me a meal, cursed them and told them
to go and play somewhere else. But they had no *bopi*; no cool,
clean streams, no vine swings. So they set off and began chasing
the dogs, goats and chickens of the nearest negro household.

At the far end of the little village stood Ekianga's house. The
only one that was not falling to pieces. It straddled the width
of the village, bridging the space between the two not very
straight rows of smaller huts. The *elima* girls made a couple of
sorties from the single door, dashing around to the back of the
house where they could not be seen. Otherwise all was quiet.

The evening meal was a good one, as the pygmies had
brought in far more meat than they allowed to the villagers;
and for what they had given the villagers they had received
generous portions of rice, beans, groundnuts and manioc in
return, and a few of the tiny, bitter tomatoes which blend so
well with manioc leaves and groundnuts in the making of a
sauce. When it was dark, the women and children retired
almost immediately, and the men gathered at the central fire
where four upright poles indicated the pygmy version of a
baraza, or men's meeting place. At first, several villagers came
to see what was going on, and the pygmies sang and danced for
them. They sang and danced until the villagers, tired after a
heavy day's work in the plantations, went to bed. Then the
men really began to sing. They sang and sang until it seemed
almost as though we were back in the forest. But any time
there was a strange sound, and it was suspected that there
might be a villager lurking about, they stopped. They had
guards posted all around to make sure they were not caught
unawares.

It must have been about three o'clock in the morning when
the *molimo* made its appearance. It came into the camp preceded by two youths, each holding glowing firebrands in their
hands. Waving the brands high in the air, then swinging them
down to touch the ground, they laid a fiery path for the *molimo*
to follow. It was stately and imposing. Following this path of
glowing cinders the *molimo* circled the village several times and
then came to rest by the fire, where it sang very quietly. Just
as I wondered if it was really going to sing as it did in the forest
it moved off, and the tempo increased. Surrounded now by all

146

the youths the *molimo* started running around, faster and faster, making lightning swoops on the group of older men who were sitting by the fire. The rest were either dancing in a frenzy around the fire itself, as if protecting it, or else they were with the *molimo*. I heard a crunching sound as the *molimo* ploughed through a small field of sugar cane at the edge of the village, and I heard leaves being ripped off roofs and the dried mud beaten from walls.

There was a subtle change of atmosphere, and a tremendous air of expectancy. By their insistent singing the older men seemed to be trying to avert some disaster, just as the younger men did by their dancing. Moke allowed nobody to sleep or rest except for a very brief spell when he kept half a dozen men singing while the others dozed as best they could around the fire. Then he woke us up and goaded us all into singing harder than ever. The *molimo* had left, and was evidently playing havoc in the negro village. But as we sang it came sweeping back into the camp from behind Ekianga's house. The two fire-dancers in front sent showers of sparks up into the air and on to the ground, the *molimo* made straight for the central fire. It knocked us aside—even old Tungana—and slowly and deliberately it tramped through the fire, scattering it to all sides.

The men who had been sitting were on their feet at once, and hurriedly threw the scattered logs back into the centre and joined the dancers in bringing life back to the fire. But the *molimo* was insistent, and the next time all the men joined with it. Reluctantly at first, then sharing the inevitability of the occasion they too danced and stamped amongst the flaming logs until there was nothing left but glowing coals. As the pace of the dance increased the embers were scattered wider and wider, until the whole of the village seemed to be carpeted with winking red eyes. The *molimo* was once again blowing martial blasts, and the men accompanied it with a deep, rhythmic chant. All except a few of the elders followed it on its last tour of the camp, slow and majestic, tramping out the fire of life until there was only one glowing ember left. I saw Moke sadly sit down on the ground beside it, pick it up in his hands and fondle it. He took a cigarette butt from behind his ear and lit it with the coal. Then he put the coal carefully back on the ground, stood up and stepped on it until it, too, was extinguished, after which he turned his back on us all and disappeared into his hut. The *molimo*, surrounded by a triumphant entourage, now left the village and we escorted it back to the forest. Dawn was breaking, streaking the sky with gold and crimson, and as soon as we had found a suitable temporary hiding place we returned to the village.

It was in a dreadful state of chaos. Black ashes and cinders covered the ground everywhere; manioc plants were uprooted, sugar cane trampled down, chairs smashed, leaves torn off roofs and the mud from many walls crumbled on the ground. The negro villagers were full of complaints, but in the pygmy village men and women were silent, sleepily clearing up the wreckage and preparing for the midday feast.

Nobles and headmen began arriving from neighbouring villages, and as even the four poles that represented the pygmy *baraza* had been smashed to the ground, they took possession of my veranda. They had come to supervise the ritual feast which, according to their custom, marked the end of the mourning ceremonies for Balekimito. They were a little surprised to find that the *elima* had occupied the house in which Balekimito had died, but the pygmies were in no mood to argue; they just shrugged their shoulders and said it was a good house. As far as they were concerned they had finished mourning their personal loss as soon as their mother had been buried. From then onwards they had been fulfilling their duty to the forest by rejoicing with the *molimo* for the long and good life that had been granted to the old lady. They had acknowledged the gift of the fire of life from the forest, and the forest's right and power to withdraw it; they had acknowledged and accepted the fact of life and death as both being, equally, the gift of their God.

But if the villagers wanted them to have a formal feast to end the mourning, and were willing to supply the food, they did not object.

The food was cooked and laid out in the centre of my veranda, which when the sun was high was the only cool and shady place in the village. This meant that the villagers were pushed to the edge as the pygmies came in to eat. The men came in and sat down and ate without ceremony while their patrons watched hungrily. But they were not related to Bale-kimito, so they could not partake. Women, too, were excluded, according to the village rules of the game, but I saw a number of dishes quietly slipped to one side, and shortly afterwards all the women were enjoying their own feast, concealed behind the *elima* hut.

The villagers asked if everyone had ritually bathed, to wash off the taint of death, and everyone politely said yes. The villagers then asked if the pygmies had any meat to spare, and the pygmies equally politely said no. They had given it all to the *molimo*, they said, and to this there was no answer, for this was a very creditable thing to do, in village eyes. I bought a large quantity of palm wine, however, and this mollified to some extent the strained feelings of the patrons. They knew they were being deceived and cheated, yet they felt they were only doing what was right in trying to impose their beliefs and practices on the forest savages who had no culture of their own. If the forest spirits struck the pygmies down for not having heeded their patrons, for not having performed all the necessary rituals as they had been told, then it was not the fault of the villagers. They drank their fill and departed.

Once the negroes had gone we all tried to get some sleep, for it was late afternoon and we had had no rest. But for three months the men had sung and danced and feasted every night and for three months the women and children had been sent to bed early, forbidden to come out and take part in the festivities except for the dance of the old woman. But now the *molimo* was over, and the women took advantage of the fact. After all, an *elima* had begun, and that was as important to the women as the *molimo* was to the men. At dusk all the girls of the *elima* came out of their hut. Their beautiful young bodies shone and glistened with the palm oil that had been carefully rubbed into the skin, making them even more alluring to the young bachelors. And together with the older women they began to sing, in deep contralto tones. The men protested loudly that they were tired and wanted to sleep. But the women just laughed and sang all the louder. The *elima* had started in earnest.

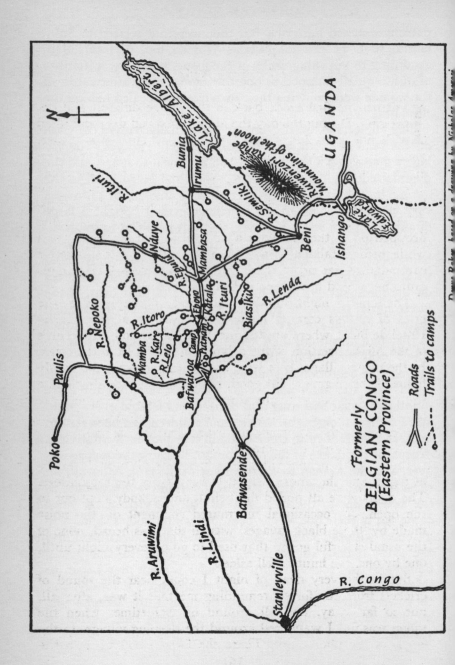

Formerly
BELGIAN CONGO
(Eastern Province)

⟶ Roads
┈┈┈o Trails to camps

UGANDA

Lake Albert

Mountains of the Moon

Ruwenzori Range

Lake Edward

R.Semliki

Bunia
Irumu
Beni
Ishango

R.Ituri

Nduye
R.Epulu
Mambasa
R.Epulu
Katala
R.Ituri
Biasiku
R.Lenda

R.Nepoko
R.Itoro
Wamba
R.Kare
R.Leio
Putnam Camp
Itoyo
Bafwakoa Camp

Paulis

Poko

R.Aruwimi
R.Lindi

Bafwasende

Stanleyville
R.Congo

Drawn Below, based on a drawing by Nicholas Amorosi

Chapter Nine

I FOUND the first week back in the village difficult and depressing. During the day the air was hot and dry, and filled with fine particles of dust. At night, when it was cooler, gnats and mosquitoes made me appreciate all the more their absence in the forest. My hut, ten times the size of my home in the hunting camp, seemed close and stuffy, and as I lay in bed I could hear the huge spiders, some several inches across, crawling about in the leaves of the roof. Occasionally one dropped on to the bed with a dull thud, and lay there for a while before stalking away. At first I carried out an active campaign to get rid of them, but it was useless—and a mosquito net would have been unbearably hot.

The familiar night sounds of the forest were replaced by the cries of drunks coming home from a dance at the nearby Hotel de Bière, where an antique Victrola ground out records of the African version of the calypso, imported from the coast. In the negro village there were constant fights—not the wordy fights of the pygmies, but good, old-fashioned fisticuffs. These were likely to occur at any time of night, depending on how late it was when an outraged husband, returning from a rendezvous with his current girl friend, reached his own hut to find his wife in bed with someone else.

In our small pygmy village the fires were all brought inside at night, and the smoke helped to keep away the mosquitoes. The doors were all pulled tight shut, and nobody slept out in the open. An occasional murmured comment on the noise made by 'those black savages' was all that was heard, none of the usual cheerful gossip that used to go on every night until, one by one, the hunters fell asleep.

But in the very dead of night I could hear the sound of crickets from the forest, reminding me that it was, after all, not so far away, and all around us. Sometimes when the moon was full I wandered around the sleeping village, to the edges of the plantations. There the forest stood up into the

night like a great black wall, encircling this small circle, cut
so ineffectively out of its very heart. It rose up all around, confident in the knowledge that it would soon claim its own and
grow back more thickly than ever, where it had been cut.
There were several such overgrown villages 'nearby not long
abandoned; yet already the forest towered above. The forest
destroyed everything but its own. Only against the machine age,
which had barely reached it, was it powerless.

In the very early hours the forest made itself felt, but close
as it was it faded far away with the coming of dawn. The
morning sounds of cocks crowing, goats bleating, and the
elephants of the *Station de Chasse* as they were put through their
exercises, all obscured the faint chirpings of crickets and birds
and the chatter of monkeys. And as the sun rose and began to
beat down on the parched and scorched earth, the village
asserted once more its determination to dominate the hostile
world around it. The men went off to cut down more trees for
more houses and bigger plantations, and the women went to
hoe and cultivate the small fields so laboriously prepared, but
which after another one or two years of hard work would
have to be abandoned. For the soil which can support the
primeval forest with abundant and luxurious ease refuses to
bear fruit to the crops of the villagers for more than three con-
secutive years.

In the neighbourhood of the Epulu there were, in fact,
several villages. Some twenty-five years earlier, when Patrick
Putnam first came to the spot and decided to settle there, there
were none. He built his home with the help of various men of
different tribes whom he took around with him to assist him in
his work. Some were from villages several hundred miles away.
They called in other workers, and gradually the village of
Camp Putnam came into being. A pretty, straggling village,
neat huts dotted haphazardly either side of a gracefully curving
track, just wide enough for a car to pass through to Pat's home.
Even though its members came from half a dozen different
tribes it quickly assumed the characteristics of a traditional,
tribal, forest village. The lack of family feeling that is so powerful where a village consists of all the men of a single clan, their
wives and children, was replaced by a common loyalty to Pat,

Nicholas Amorosi

who worked day and night in his small dispensary, saving the lives of the people he loved and, in the end, giving his own. Pat acted as Chief in nearly all respects. Disputes were brought to him and he settled them, wisely and justly, according to the tribal laws he knew so well. The village began to attract pygmies, who gradually attached themselves, again in a perfectly traditional manner.

To a lesser extent this was so even with the new and distinctly commercial village that was set up on the roadside, by the government, immediately after Pat's death. The main focus of the village was not, as in so many villages, in either special plantations or road work, but in the *Station de Chasse*. This, with its collection of live forest animals, became an attraction for wealthy tourists with time and money enough to make the trip from Stanleyville or Beni; hence an optimistic Belgian had built a large motel directly opposite. A little further up the road, away from the river, were the houses of several hundred workers. This village stretched right up to the entrance of Camp Putnam, and beyond was the Hotel de Bière and a line of a half-dozen or so stores. The majority of these villagers, as in Camp Putnam, were from distant districts and from many different tribes. But they had nobody to hold them together, as Putnam had done for his workers; their disputes were settled arbitrarily according to Belgian notions of right and wrong. It was a village filled with jealousies, deep-seated grievances and smouldering hatred.

A few kilometres to each side of the Epulu River were two traditional villages. To the west was Dar-es-Salaam. The name was brought from the coast by the slave traders, and its inhabitants were all descendants of the almost universally disliked BaNgwana, who had sold their souls to the traders and helped them in the capture of slaves and the theft of ivory. The headman of Dar-es-Salaam was a friendly rogue, Musafili. About two kilometres away from Dar-es-Salaam was the camp of Musafili's pygmies. Like every other negro he referred to 'his' pygmies as though they were so many chattels. Although Dar-es-Salaam always offered the warmest hospitality, I never felt at ease there. The heritage of a bloody and treacherous past was only barely dormant, and the BaNgwana are suspected today, as they always have been, of practising the most dreaded forms of witchcraft and sorcery.

A little further to the other side of the Epulu was a very different village, and much more typical of the traditional settlement in this area. It was named Eboyo, and the villagers belonged to the BaBira tribe, predominant in the southern and central Ituri. Here there were two chiefs; the traditional chief, Sabani, and the government-appointed chief, Kapamba. Kapamba's job was to see that the right crops were planted, harvested, and marketed to the Belgian Government, and that the taxes were collected and paid over. The Belgian policy was largely to let the people in this rather difficult and unprofitable part of the Congo live as they had always lived. The only stipulations were that they should live peacefully, and that in addition to their own small plantations they should cultivate more ground for crops which would both help to feed immigrant labour forces, such as the road gangs and the *Station de Chasse*, and would also be of commercial value to the Belgians in the world market for such products as cotton, palm oil, and groundnuts.

Kapamba was not a popular figure. He was small, pathetically ugly, and his hut was at the very far end of the village, behind a large hillock on the top of which stood his *baraza*. Only his close relatives and a few pygmies lived near him.

In the village proper, Sabani's house was in the centre. Sabani was a sly old man whose father had married a Lese girl, from the eastern edge of the forest. Sabani changed his tribal affiliation to suit the convenience of the moment, and the Belgians found it impossible to deal with him, so they appointed Kapamba in his place. But Sabani was still the real power in the village, and the real authority. He was also the chief ritual authority for miles around, and he was a frightening sight when painted and dressed up for his role of witch-doctor at the time of the *nkumbi* initiation ceremony. Even on other occasions, when he wished to be formal—such as the visit of a Belgian Administrator—he was no less inspiring. Ragged trousers came down to within a few inches of feet that were barely covered by ill-fitting shoes, slit where they pinched so as to be more comfortable. An old khaki tunic covered his bare chest, and on his head he wore either a pith helmet decorated with gennet skin, or else a long-haired wig of monkey fur. In his hand he carried a metal walking-stick that had once been the handle and stem of an umbrella. Sometimes he wore the gennet skin around his waist, or tucked into his belt, or hung from his neck. It was the symbol of his ritual authority, and it contained the charms from which he derived his power. Generally Sabani was a kindly and gentle man, wise and good. But his loyalties were narrow, confined to his own kin and a few friends. Beyond that circle he was a man to be feared rather than trusted.

Eboyo also had its pygmy village, not far from the main village, but Sabani preferred to keep his pygmies in the forest where the administration could not see what they were doing. Generally they were tending illicit plantations, of hemp amongst other things. Sabani also had his own still, where he used to have a powerful spirit made from manioc. He was frequently drunk but never incapable. At all times he was an imposing and commanding figure. In his village there were some thirty families, including Kapamba. Camp Putnam and Dar-es-Salaam were about half the size.

The attitude of all these villagers to the forest was the same. They made their villages as open as possible, and they built their houses without any windows. They tried to ignore the world around them, because for them the forest was hostile, something to be feared and fought. Even after cutting it down there was the constant labour required to keep it from growing back. It was filled with evil spirits that cursed the soil so that although it would bear gigantic mahogany trees it yielded only the most meagre fruits for the villagers. The tribes of this area had once been plains dwellers, with herds of cattle, living in the friendly, open spaces of East Africa. When they were driven into the forest by more powerful tribes, they brought as much of their plains culture with them as they could. But their cattle died off, and they had to live by cultivation alone. They brought with them their ancestral beliefs, but the spirits of their ancestors resented being forced to live in the forest away from their rolling, open homelands. In the plains the very ground was sacred to the ancestors who had lived there for generations, but here there could be no such bond. The graves themselves had to be abandoned to the forest sooner or later, as the clans shifted endlessly in search of fresh soil.

While the only place the villager could call home was the temporary, sun-scorched clearing which he had cut for himself, and in which he would live a few years at the most, to the pygmy the whole forest was home and the village an isolated and foreign territory; hostile but ultimately impotent, to be regarded with indifference, amusement, and occasionally with annoyance. The village people were envious of the pygmies, though they tried to look down on them. But try as they might, they were compelled to respect them because of their own belief in the mystical bond between a people and the land of their ancestors. They had no such bond. Just as they were afraid of the spirits of the forest, so they were secretly afraid of the pygmies, and made them offerings of the first fruits of their plantations, and gave them an important part in the ceremonies for the village dead—the dead who had to be buried in foreign soil that might belong to the village world now, but which would soon belong once again to the forest.

The relationship between the two people was a strange one, full of ambivalence and uncertainty. By and large the village was their only common meeting place, for the negroes disliked going into the forest except when absolutely necessary. And the village was the world of the negro, so he assumed a natural position of authority and domination in that world. But it was a position without any foundations, for the pygmy only had to take a few steps to be in his own world. The negro was unable to exercise any physical force to maintain his control over the pygmies, so he created and maintained the myth that there was a hereditary relationship between individual pygmies and families and individual negroes and families.

It was not a question of slavery, rather of mutual convenience. The cost to the negro was often high, but it was worth while. The pygmy was able to do all the necessary chores for him that necessitated going into the forest, and could bring him game that supplemented an otherwise largely vegetarian diet. And in latter days, with the compulsory maintenance of larger plantations, the negro found the pygmy a useful source of additional labour. But for all of this he had to pay, and he was always uncertain as to how long any one pygmy or group of pygmies would stay with him. They were frequently miles away in the heart of the forest, and if the hunting was good no amount of coaxing would bring them back.

For the first week back at Camp Putnam, then, I found the villagers going out of their way to be generous to 'their' pygmies. The pygmies wandered about idly, hanging around wherever they found food cooking, occasionally responding to a request to help with some kind of work. The village black-smith, a man of many parts, adopted Masisi's family, though Masisi himself would have nothing to do with him. He said that Baziani was a trouble-maker, which he was, and that he was after Masisi's daughter, which was also true. But Baziani was wealthy, and Masisi's children spent most of their time in his *baraza*. They were well fed and looked after, and treated more as friends than anything else. Baziani, in return, made use of them to run errands, usually connected with his virtual monopoly of the local hemp market.

Baziani was genuinely fond of the pygmies, and in the evenings he used to bring out his drums and the pygmies gathered around and danced. Then after a while Baziani would relinquish his part as a drummer, acknowledging the superiority of his friends. Then he sat back in his cane chair and watched with a contented smile. He was one of the first to help in time of need, and so it was that he was one of the few villagers who was never short of meat, or plenty of leaves for the roof of his house, and he invariably had a crowd of good-humoured, well-fed pygmies sitting around in his *baraza*, waiting to do him some small favour.

Masisi, like Njobo and many of the others, had for a long time been attached to a noble of the BaBira tribe, Ngoma, uncle of my erstwhile, would-be wife, Amina. But Ngoma was wealthy, and had many other pygmies so he did not resent the defection of a number of them to Camp Putnam. In fact he frequently came to visit them, or sent messages to find out how they were, and when they were near Ngoma's home, which was some distance away, they were presented with stalks of plan-tains and baskets full of groundnuts and manioc. Whenever work on Ngoma's plantation was behind schedule he could be sure of willing help that would avert the penalty of a heavy fine or confiscation.

Manyalibo, however, was attached to Kapamba, the government chief at Eboyo. For all his outward clowning, Manyalibo was crafty, and he played Kapamba for what he could get. Eboyo had some of the best plantains in the region, and it was much closer than Ngoma's village. He seldom went there himself, and he refused all calls by Kapamba to come and help in the plantation. Sometimes he sent one of his daughters, or his adopted son, Madyadya, to help the chief and to maintain some semblance of a bargain, but they never stayed for long. Once, at Apa Lelo, Kapamba was in such dire need of help that he came out to the hunting camp himself. Manyalibo received him courteously enough, and housed him for a few days, but firmly refused to return to the village with him or to allow any of his family to return. Kapamba alternately flew into tantrums and attempted to bribe Manyalibo with stories of the abundance of rice and plantains waiting for him at the village. Finally Manyalibo agreed to let his wife go, and he sent her off carrying a basket with two legs of antelope as a gift for Kapamba's wife. The camp watched the chief leave with few regrets. They had allowed him to stay because he had brought a quantity of food with him.

Manyalibo's wife returned two days later with her basket full to overflowing with delicacies from the village. She had told Kapamba that she did not want to stay longer without her husband, so he had no option but to allow her to return.

Manyalibo's son-in-law, Ausu, was even more of a renegade. His 'hereditary master' was André, the chief of a powerful Islamic village just beyond Eboyo. André's father, old Effundi Somali, had been a child when Stanley brought his relief expedition through the forest to rescue Emin Pasha. He used to tell stories of the dreadful wars that were fought in those days, and of the trail of destruction that Stanley had left behind him. While the old man was still alive André had worked as head cook at Camp Putnam, but now he was chief in his father's place.

Part Bira and part Ngwana, André inherited all the traits of the latter. In his dealings with Europeans he could be charming and helpful; he could also be cunning and crooked. With other villagers he was proud and avaricious, and he had a supreme contempt for the pygmies—which he could disguise whenever it suited him.

A few years earlier, when Ausu had been hunting forest pig, a boar had charged him and wounded him badly, ripping his stomach and one leg. André had taken Ausu into his own house and cared for him, nursing him for months until he was well again. This was done under the guise of friendship, and I have heard André tell the story many times to illustrate how enlightened he is. But I have also heard Ausu's comment that André had done no more for him than he would have done for a valuable hunting dog. A pygmy is not unresponsive to genuine kindness, and Ausu would not have left André soon after recovering if there had been any such kindness in the act. But he left, and refused to return under any circumstances.

A few years later I was with Ausu in the forest when we were returning from a hunt. We were up above the Lelo River, where the forest was clear of all undergrowth. Ausu saw something move amongst a heap of leaves, and we found a baby okapi lying there. The mother okapis leave their young, covering them with leaves, when they go in search of food. Knowing that the *Station de Chasse* was anxious to have a small live okapi Ausu picked this one up and brought it back to camp. We took turns in carrying it over our shoulders as although it was only a few days old it was both heavy and frisky.

We eventually got it safely to the *Station de Chasse*, and anticipating some kind of trouble I asked the director to arrange to pay the 500 francs directly to Ausu. This was the standard sum offered as a reward for a live okapi, but only if it survived a week. We returned to the forest, and Ausu regaled the camp with stories of what he was going to do with the money, the equivalent of about ten dollars. Manyalibo asked him if he was going to give any to André, and this raised a storm of protest. André was universally disliked by the pygmies. But Ausu said that he would give him some, and eventually it was decided to give the chief 200 francs out of the 500.

The very next day a delegation arrived from André's

village, with a letter from André to me saying that he had heard Ausu was to be paid 500 francs, and would I see that it was all paid over to him as Ausu was his pygmy and had refused to work for him for several years. I sent a reply back saying that the *Station de Chasse* made their payments direct to the person who caught the animal, and that Ausu had offered to pay André 200 francs. For the next few days we had a series of notes and delegations, pleading, cajoling and threatening. Finally André said that if Ausu did not give him all the money that he was going to bring a charge of theft against him and have him put in jail. As the tribunals are all negro, with no pygmy members, even the most obviously distorted charge against a pygmy was bound to succeed.

Ausu decided that this being the case he was not going to give André a single franc. On the day of the payment I went with him to the *Station de Chasse*, saw him get the 500 francs, and joined the director in advising him to climb over the fence and return through the forest rather than leave by the main gate. And sure enough, as I left, there was André with half a dozen of his men, waiting. When he saw he had been tricked he went to the paramount chief of the area and had a court order made out for Ausu's immediate arrest and confinement to prison. But Ausu was already back in the forest, and in time even court orders are forgotten. And so with all his show of authority and superiority André was unable to do anything. Trying to pursue a pygmy in the forest was a task no villager would undertake for any amount of pay.

There are those, like André, who really believe that the pygmies are inferior and are meant to be treated like slaves, but for the most part the villagers are much more sensible and realistic. If they want to have control over their pygmies, they know it must be based on something other than threats of physical force. But even the most benevolent of them regard the pygmies much as the pygmies regard them in turn, as a kind of convenience. I only know of one villager who was almost accepted as one of themselves by the pygmies. That was Kaweki, the fisherman.

Kaweki was a not-so-young bachelor of the BaBira tribe. He claimed to dislike women, saying they made far too much noise and cost far too much money. Most of the time he was in the forest, catching the fish in dams and traps, drying them and sending them down to market with his pygmy friends. This struck the other villagers as an undignified, not to say an unsafe, procedure. But Kaweki was not one to care about what others thought.

After some persuasion he agreed to allow me to come out to his fishing camp. He was known to dislike and to actively discourage visitors, so I was not surprised when he was vague as to the camp's whereabouts, merely saying that his friends would bring me there the following week. His 'friends' turned out to be Kelemoke and Hari, a young pygmy from a village the far side of Ngoma's. Hari was a cousin of Kenge. The four of us set off before dawn one morning.

Even when a whole camp is on the move, men, women and children, the pace of the pygmy is swift. But when there are just two or three active young hunters they maintain a steady trot for however far they have to go. I found myself running to keep up, and we only stopped twice during the entire journey. The first time was when we were not far from the Lelo River. We rested among the ruins of an old, deserted camp, and we each ate a banana and smoked a cigarette. The second time was shortly after midday, when we met Moke and another on their way back from Kaweki's camp. They were surprised that we had come so far; they themselves intended to stop over in one of the old camps for the night. Each of them carried a smouldering ember wrapped in a leaf, in one hand, and a bow and bunch of arrows in the other. They warned us that Kaweki was not well, and said that it was the '*dawa nde BaNgwana*'—the magic of the BaNgwana. This made Kenge, Hari and Kelemoke all cluck their tongues, because the *dawa* of the BaNgwana is the most powerful of all. Hari said that he had heard that Musafili was jealous because Kaweki was making so much money with his fish; Kenge, however, said that he thought it was Kaweki's own people, the BaBira of Babama village. They were offended, he said, that Kaweki should live in the forest with the BaMbuti instead of in his own village. They suspected him of being in league with the evil spirits of the forest. Kelemoke nodded and said yes, it was true that Kaweki fell ill every time he returned to the village, and that it must be the strongest kind of witchcraft to reach so far out into the forest. Everybody was unhappy, and we resumed our journey at a slower pace.

We were in an area where there were a number of streams, tributaries to three large rivers which had their source not far to the north-east, spreading out in different directions. There was a lot of heavy undergrowth and we had to follow trails made by antelope. Even the pygmies had to bend almost double. Hari was leading, and Kenge shouted at him from behind that there was a much better and quicker way. Hari laughed at his cousin and said, jokingly, that Kenge was a white man's pygmy, how could he know anything about the forest? Kenge made some uncomplimentary retort, and motioned for me to follow him. We ducked off to one side, and for a while the going was worse than ever. Very quickly the sounds of the other two were lost, and I had momentary doubts for every now and again Kenge stopped to listen, peering this direction and that. After another couple of hours Kenge pointed ahead and I heard a new sound, a distant waterfall. '*Apa nde Kaweki*,' he said; 'Kaweki's camp.' Usually he would have said a good deal more, but I was glad to see that the long day's trek had left even Kenge breathless.

We came to the narrow but swift River Kare, and waded through it with caution. Kenge said it was a dangerous river, unexpectedly deep in places, and full of the small but vicious forest crocodile. It was here that Kolongo, Ekianga's other brother, had died two years ago. The waters swirled angrily; they were muddy and full of debris, the banks were slippery, and it was plain there must have been a heavy rainfall recently. On the far side the ground rose sharply before dipping down to the valley where the Itoro River flowed with a grandeur only found with forest rivers. The undergrowth disappeared, and once again we were in one of those sylvan glades where every leaf seemed to tremble a welcome. It was warm and friendly, and as we began to descend the rumble of the waterfall swelled into a roar.

Kenge made his way to the edge of the Itoro, just below the falls; it was wide and strangely smooth, but flowing in spate with all the majesty and power of a river twice the size. Kenge said we would wait for the others there, so as not to shame them in front of Kaweki by arriving first. He went to the water and washed himself. He disapproved of my refusal to do the same, but I was too tired even to sit down. I just propped myself up against a tree and waited. It was not much more than fifteen or twenty minutes later when Kelemoke and Hari came into sight, the one trotting quietly behind the other, apparently as fresh as when they had set out. They feigned surprise at seeing us, and said they had stopped to hunt on the way, and had thought we should be in the camp by now. Kenge retorted that he was sorry they had caught nothing—we had been waiting two hours, maybe three or four, in case they were lost. We moved on, and in a few minutes, by-passing the falls, we came once more to the banks of the Itoro, and into Kaweki's camp.

It was a mixture of negro village and pygmy camp. The clearing was only some twenty yards across; on one side was Kaweki's house, built in the rectangular shape of a typical village house, but with walls hung with leaves instead of packed with baked mud. At right angles to it was another hut, long and low, the eaves coming right down to the ground so that the entrance gaped in a great triangle. We dropped our bundles on the ground and went into Kaweki's hut, calling from outside that we had arrived. The hut was divided into two by a leaf screen, the outer half being the kitchen and drying rack for the fish. The rack was full, and a fire underneath did double duty, cooking food and smoking the fish. On the far side of the screen Kaweki lay on a bed of fresh leaves. A young pygmy girl knelt beside him, gently massaging his chest. Hari greeted his sister, then went to Kaweki, and addressing him as 'father' asked him how he was. Kaweki was almost too weak to speak, but he took our hands in turn and apologized for not having been able to get up to greet us. He said, 'My daughter will get you something to eat.' He gave Hari's sister a playful push on the head, then lay back and seemed to go to sleep.

Hari led us out and into the long, triangular hut. We went right through to the back and there, on each side, was a narrow bed of sticks where he said Kenge and I were to sleep. This was the visitor's hut, he explained. He and his sister slept in Kaweki's house. I asked who the visitors were, as Kaweki was reputed to have none. 'Oh, I don't mean villagers,' Hari said, 'I mean pygmies, his friends.'

I had barely washed and sat down in the middle of the clearing to eat the food prepared by Hari's sister when there was the unmistakable sound of a band of pygmies approaching. In a few minutes a score of them burst into the clearing; young men and women, laughing and shouting their greetings.

They said that they had heard that Kaweki was ill, so they had come to hunt for him and make him well. What had promised to be a peaceful rest in a lazy fishing camp now threatened to be something very different. The pygmies had come as far as we had, but within minutes of arrival they were off again for a quick hunt before the sun set. Kenge himself joined them, taking Kaweki's two hunting dogs and arming himself with a stout, short-shafted spear. The women stayed behind to improvise some shelter for the night. But they decided it was too much work, so they just added some sticks and leaves to the visitor's hut, forming an elongated windbreak that reached almost over to Kaweki's house. By the time the men came back, empty-handed, they had installed themselves and were boiling pots of plantains and manioc over the fire, together with fish to which they had helped themselves from Kaweki's rack. That night those who could not squeeze into the shelters or into the overcrowded visitor's hut slept in Kaweki's house. I myself found that the longed-for blessing of sleep did not come easily. Kenge had, of course, found an old girl friend amongst the newcomers, and the two of them lay together, beside me, all night long, flirting and laughing. Even though I lost a lot of sleep I got a new look at the facts of life.

Over the next few days I saw the hunters go off to bring back the meat that would make their friend well and strong; and each day they collected roots and herbs which they prepared and gave to Hari's sister to treat Kaweki with. Just their very presence, boisterous and cheerful, was a tonic to the invalid. On the second night he came out into the clearing and lay down beside the fire, covered with a blanket. He joined in the conversation of the pygmies, joking and telling stories in their own manner. Then he half-sang and half-chanted legends of the past, urged on by Hari who sat nearby with all the attentiveness and affection of a son, smiling happily when he saw the villager that he loved as a father slowly regain his strength and life.

In a few days Kaweki was up and about, though he still tired quickly. He thanked his friends and told them that they had no need to stay longer if they wanted to go. But it was a happy camp, and everyone was in favour of staying. Kaweki had built a dam above the falls, channelling all the fish into a long, wicker trap, some twenty feet of elaborate basketry which caught the fish and held them firmly in its meshes until they were pulled out and killed. But the recent storm, which had broken just the day before we arrived, had breached the dam and damaged the trap. The pygmies decided that they would rebuild it for Kaweki, and while a few of them hunted the rest went off and cut the wood needed, and returned to work under Kaweki's supervision.

In the evenings, the pygmies sang songs they would ordinarily never have sung in the presence of a villager. But Kaweki was a lover of the forest—and to the pygmies that was all that mattered. And on the evening when Kaweki actually danced and declared himself completely cured, the pygmies responded by singing a song of the *molimo*, just as they would have done for one of their own people; a song of praise and thanks to the forest for saving one of its children, even if he was a villager.

Chapter Ten

THE PYGMIES and the villagers came into contact with each other, both as individuals and in groups, and it was inevitable that over the years they should each adopt some of the others' customs. The contact between them was sometimes intimate and friendly, as with Kaweki; and sometimes it was formal and hostile, as between André and Ausu. Most frequently its purpose was just a matter of barter, and at all times each regarded the territory of the other as a foreign place.

When customs were adopted by one people, they gave it their own particular character and meaning, and frequently changed it so that it was all but unrecognizable, other than through the common use of the original name. Such was the case with the *molimo*, and such was also the case with the *elima*. Who borrowed from whom is a difficult question, and perhaps not a very important one. The difference that emerges in the process of borrowing is much easier to see, and adds much more to our understanding.

To both people the *elima* is concerned with the arrival of a young girl at the age of maturity, her transition from girlhood to the full flowering of womanhood. This coming-of-age is marked more dramatically for girls than with boys by the physical changes that take place. Both pygmies and villagers recognize the significance of the first appearance of menstrual blood, and both celebrate it with a festival which they call '*elima*'. But here the similarity ends.

My ex-girl friend, Amina, told me a great deal about the *elima* among the BaBira, and I learned more from older women in different villages that I lived in from time to time. Everywhere, right through to the very edge of the forest, I found the same story among the villagers. Blood of any kind is a terrible and powerful thing, associated with injury and sickness and death. Menstrual blood is even more terrible because of its mysterious and constant recurrence. Its first appearance is considered, by the villagers, as a calamity—an evil omen. The girl who is defiled by it for the first time is herself in danger, and even more important she has placed the whole family and clan in danger. She is promptly secluded, and only her mother (and I suspect one or two other close and senior female relatives) may see her and care for her. She has to be cleansed and purified, and the clan itself has to be protected by ritual propitiation from the evil she has brought upon them. At the best, the unfortunate girl is considered as a considerable nuisance and expense.

It is generally assumed that the girl was brought into this condition by some kind of illicit intercourse, and her mother demands of her who is responsible. This is the girl's chance to name the boy of her choice, whether or not she has even so much as spoken to him. He is then accused, and has to make suitable offerings which are not only necessary to help in the protection of the girl and her family, but also for his protection, since he had been linked with her verbally, however untruthfully. He may then deny the charge, and there may be some litigation if the girl's parents think he would make a suitable husband. Or else they may consider him unsuitable, and let the matter drop. If he likes the girl, and accepts the responsibility, this constitutes the first step towards official betrothal. If he disclaims all responsibility, then the girl has to try again, as until she has a husband the danger is thought to remain.

The period of seclusion varies from tribe to tribe, and even from village to village. Sometimes it is just for a week or two, sometimes it lasts a month or more. And sometimes it lasts until the girl is betrothed and can be led from her room of shame to be taken away by her husband. At the final wedding ritual she is ceremonially cut off from her family, and told that she is now her husband's responsibility and must fend for herself and not come running home every time she gets beaten. The point being, of course, that if she does this then her husband can claim back the wealth he will have already paid over in the expectancy of her fidelity and success as a wife and bearer of children. And by the time the disaster occurs her family will almost certainly have spent the money to obtain a bride for one of her brothers.

The whole affair is a rather shameful one, in the eyes of the villagers, as well as a dangerous one. It is something best concealed and not talked about in public. The girl is an object of suspicion, scorn, repulsion, and anger. It is not a happy coming-of-age.

For the pygmies, the people of the forest, it is a very different thing.

To them blood is, in the usual context in which they see it, equally dreadful. But they recognize it not only as being the symbol of death, but also of life. And menstrual blood to them means life. Even between a husband and wife it is not a frightening thing, though there are certain restrictions connected with it. In fact they consider that any couple that really wants to have children should 'sleep with the moon'.

So when a young pygmy girl begins to flower into maturity, and blood comes to her for the first time, it comes to her as a gift, received with gratitude and rejoicing; rejoicing that the girl is now a potential mother, that she can now proudly and rightfully take a husband. There is not a word of fear or superstition, and everyone is told the good news.

The girl enters seclusion, but not the seclusion of the village girl. She takes with her all her young friends, those who have not yet reached maturity, and some older ones.

In the house of the *elima* the girls celebrate the happy event together. Together they are taught the arts and crafts of motherhood by an old and respected relative. They learn not only how to live like adults, but how to sing the songs of adult women. Day after day, night after night, the *elima* house resounds with the throaty contralto of the older women and the high, piping voices of the youngest. It is a time of gladness and happiness, not for the women alone but for the whole people. Pygmies from all around come to pay their respects, the young men standing or sitting about outside the *elima* house in the hopes of a glimpse of the young beauties inside. And there are special *elima* songs which they sing to each other, the girls singing a light, cascading melody in intricate harmony, the men replying with a rich, vital chorus. For the pygmies the *elima* is one of the happiest, most joyful occasions in their lives.

And so it was with happiness that we all heard that not one, but two girls in our camp had been blessed by the moon.

At first it was just Akidinimba, daughter of Manyalibo's sister. Akidinimba was a notorious flirt who considered herself the most beautiful girl in the world. She was just sufficiently plump to be envied by less well-endowed girls, but not so plump that her maidenly figure was lost. She was particularly proud of her breasts, which were certainly the largest in that part of the forest, and she showed them off to advantage, when dancing, by adopting an unusually springy gait. As her breasts bounced up and down the eyes of all the bachelors followed their movement with pleasure. Everyone agreed that Akidinimba's *elima* would be a lively one.

But Manyalibo's niece was in no hurry, so for the first few days she remained in the hut of her cousin, Kondabate, and there were no particular signs of activity visible from the outside. Then we heard that Kidaya had also been blessed with the blood, and she went to join Akidinimba. Kidaya was a niece of Ekianga, and she was more modest than the rather shameless Akidinimba. Kidaya was more handsome than she was pretty, and she had a reputation for an ability to beat up even the strongest of youths if their advances were not welcome to her. So the *elima* promised to be exciting as well as lively.

The two girls each chose their personal assistants, and called in all their young friends to join them in the hut of Kondabate and her husband, Ausu. Kondabate built on a special extension to accommodate them. This was in Apa Kadi-ketu, and it was not long after that they moved to the village, where they took over Ekianga's house. Mambunia, Masisi's brother, went with them to be 'father of the *elima*', to make sure that they were protected and properly cared for. Asofalinda, Kidaya's aunt, went as 'mother of the *elima*', to look after the personal needs of the girls, to see that they got the right food, and generally to act as chaperon. Mambunia was a bachelor, and Asofalinda a widow.

When we arrived at the village for the close of the *molimo* festival, the *elima* was well under way. The girls had already learned most of their songs, and they were now at the stage where they were taking a more active, even an aggressive interest in the eligible bachelors of the neighbourhood. In the evenings, when they assembled outside their hut to sing, they were carefully guarded by their mothers and older sisters. But the girls and their attendants cast sly glances at the young men who always drew close to watch and listen. They whispered and laughed among themselves, and when they stood up to dance they often made it very clear by their gestures which boy had taken their fancy. Sometimes they chose at random simply to annoy their special lover, and boys often stalked away in a huff. That was a mistake.

At any time of the day the girls were likely to emerge from the *elima* house, armed with long *fito* whips. If you were close enough you could hear them plotting their strategy behind the door, but there was no escape once they burst out into the open. Any male, young or old, and particularly those who had shown annoyance at being teased, were liable to be chased and whipped by the eager young furies. I myself was chased once, for a good half-mile down a forest trail, running as fast as I could. It was considered cowardly to run; but I ran with good reason, for to be whipped establishes an obligation to visit your assailant in the *elima* house. Once inside there is no need to do anything further, but you are subject to considerable attention if you refuse. And to get into the house, in the first place, requires no small amount of courage and strength.

So I ran. Most other bachelors had less reason and no desire to escape, so they contented themselves with giving as good as they got, throwing sticks and stones at the girls, running around the village, dodging behind houses and through fields of sugar cane. The battles never lasted long, but sometimes they came to an abrupt end. The girls often whipped old men or young children, and it was generally considered a compliment in either case, and did not carry the same obligations that it would have done had they been eligible bachelors. But sometimes they hit too hard, and the old men resented this and shouted, while the young boys simply screamed their heads off. In the one case Mambunia would come out of his hut in a temper and send all the girls back to their house, and in the other case Asofalinda ran amok amongst the girls, boxing their ears until they, in turn, cried like the children they had beaten.

But the *elima* beauties did not confine themselves to their own village. After all, most of the men there were relatives, and could not decently be invited into the *elima* house. Many strangers came to visit from other groups, but sometimes the attendance was disappointing. Then the girls would set out one morning to visit one of their hunting camps. I followed them on such an escapade, having first secured a promise that I would be given complete immunity. It was a strange procession, the girls all oiled and painted to look their best, wearing grass skirts in imitation of the boy's initiation (an innovation that infuriated the villagers who felt that their customs were being desecrated, and it caused open hostilities towards the end), and carrying extra stout whips, twice as tall as they were. There were about eight of them in all, led, of course, by the irrepressible Akidinimba, bouncing her breasts up and down in a determined manner as she marched along the road, past the *Station de Chasse*, and over the bridge to the far side of the Epulu River. The villagers came out to look as we passed, but from a respectable distance. On other occasions I have seen a party such as this dive into the midst of a group of village men and women, lashing out with all their might—not entirely in the friendly way in which they would whip, say, old Tungana as a compliment to his virility.

On the far side of the Epulu we climbed up the hill towards Eboyo, the sun beating down on us so that the chatter and laughter of the girls was almost stilled. But just before reaching Eboyo we turned off, following a trail through an old plantation and into the forest. As the trees closed in around and above us, the girls began to sing. They sang loudly and proudly, marching with even more determination than before, their heads back, looking at the familiar leafy world that was theirs. One at a time they sang ribald little solos, each trying to outdo the other, the rest joining in with a throaty chorus that cascaded downwards like a waterfall. Each girl held her note as the others came in after her, swelling the volume of a great chord of music until they had no breath left. Then they all stopped together, sharply and suddenly, and cocked their heads to one side to listen for the echo sent back to them by the forest.

And so we continued our noisy, festive way, until we had

climbed up and down the forest paths for about an hour. On this side of the Epulu the forest was hilly, and at times the path was steep and difficult. But still the girls sang. At last we came to the edge of one of Sabani's hidden plantations, and the girls became silent. They made their way cautiously to the edge of a cluster of houses built in the middle of the plantation. This was where some of Sabani's relatives lived when they were keeping an eye on planting or harvesting activities. Akidinimba explained, in a gusty, confidential whisper, that her very special boy friend, Pumba, would probably be there. She was very much in love with him, she said, far more so than with any of the other boys. And she wanted him. The only trouble was that he did not seem to want her.

Having discovered that Pumba was indeed there, dozing on a chair in the *baraza* of one of the houses, the girls entered the village with a rush, shouting and yelling and brandishing their whips. Pumba awoke with a start, but he was too low down in the chair and could not get up in time to escape. He was plainly terrified because in her determination to get her man, Akidinimba had detailed off all the other girls to go for him alone. When Pumba finally managed to break away and run blindly into the safety of the hut, slamming the door shut behind him his body was covered with long, angry welts, some of which were bleeding. Akidinimba wiped her nose with the back of her arm, and smiled grimly. She was well pleased with her success, and simply sat down outside the door waiting for Pumba to emerge. Pumba knew that he was no match for a girl with such patience and perseverance. He opened negotiations from the other side of the locked and bolted door.

Akidinimba was in no hurry, so she carefully examined the door and the single window, to see if there was any chance of breaking in. But the door was stout, and the window was tightly closed by a wooden shutter. So she decided, with some statesmanship, to negotiate. In return for a promise that Pumba would be allowed to emerge free of any danger of further attack, and free of any other obligations, she extracted a reluctant promise that he would attend the final stages of the *elima* back at Camp Putnam. A very downcast Pumba emerged, and immediately started complaining that there had really been no need to beat him quite so hard. The girls asked him where the other pygmies were, and he pointed across the plantation to the far side, where the forest stood out darkly into the blue sky. The girls told him to stay behind, and not to try and warn his friends. They then set off, this time with Kidaya to the fore, picking their way across the tangle of fallen trees slowly and silently.

On the far side we entered the forest and continued southwards for a short while. Kidaya, like a general plotting her strategy, sent out a couple of scouts. They came back and taking their advice we veered around to one side until we came to a slight rise. From the top of this we looked down through the trees across a leafy floor, completely bare of any undergrowth except for a few ferns. Below us lay a typically picturesque pygmy camp, shimmering in the haze of smoke from innumerable fires. It was built in two circles, joining each other as if in an attempt to become one. We could hear the idle chatter of men and women as they lay on their beds or on their hunting nets, dozing in the midday warmth. A dog wandered lazily from hut to hut, sniffing hungrily, confident that his masters were far too sleepy to bother him. The sun, directly overhead, came down through the trees in irregular, jagged shafts, so that parts of the camp were bright, others in almost total darkness. It must have been quite an old camp, for all the leaves of the huts were dry and brown.

We crept right up to the nearest hut without being seen, and then Kidaya ran lightly across the clearing and gave the sleeping youth of her choice a sound thwack with her whip. He leapt up, not so much in fear or anger as in surprise, and in shame that he should have been caught. He grabbed for his bow and picked up a small stone which he fired back at Kidaya as he ran away. Kidaya laughed with delight. This was a sure sign that her overtures had been accepted.

The whole of the camp was awake before the rest of the girls had had time to whip more than a few boys, and it soon became alive with lashing whips and flying sticks and stones. The women and children crept into the backs or their huts for shelter, while the men ran from rubbish dump to rubbish dump, gathering the banana skins they used as ammunition.

When they could not find any, they used whatever came to hand. A slice of banana skin fired from a taut bow at close quarters can be quite painful, and the girls found that the boys were putting up a good fight. So they picked up a few vine baskets that had been left lying around, and using these as shields they advanced on their quarry.

The fight got more and more fierce, and at one time the boys managed to drive the girls right back into the huts. Then they stood at the entrances and fired their pellets of banana skin inside, giving rise to squeals of anguish. But Kidaya was too tough to be done down so easily. She emerged from her shelter without even a basket for protection, and made a dash for the nearest fire. As she bent down to pick up a burning log, her particular boy friend fired a pellet that struck her on the buttocks with a smack. Kidaya spun around and hurled the log with all her might—then she rubbed her backside affectionately and smiled. The other girls followed her example, and soon there was hardly a fire left in the camp, and burning logs were scattered everywhere. The men retreated to the outskirts of the clearing and stood uncertainly, peering around trees or over the tops of huts, their bows and banana skins at the ready.

But the girls felt they had achieved their end. Each one had marked her victim, who was bound by the strongest bonds of self respect to respond to the challenge and brave the *elima* house itself. Pumba, who had followed at a safe distance, sat disconsolately on a log that Akidinimba had thrown at him in a friendly way, and he did not even look up as we left.

The girls sang all the way home; even when they left the forest and made their way back across the Epulu, gleaming orange and gold in the late afternoon light, they sang. They sang so that everyone should know that they were the *Bamelima*, the people of the *elima*, girls who had been blessed with the blood and were now women.

There were occasions when the mood was different. One evening in particular, after a relatively quiet day, nearly all the women of the group gathered together outside the *elima* house. It was towards the end of the time that had been set, in an arbitrary fashion, for the duration of the festival. The pygmies felt they had been in the village long enough, and the *elima* was interfering with their hunting. As much as the young men enjoyed hunting, the attraction of the *elima* house offered strong competition. If the boys were away too long, due to pressure from their parents, then the girls complained that they were not being treated properly, and everyone agreed that this was an insoluble conflict, unless the *elima* were either brought to an end or moved out to the forest. It was not only the boys that were away from the hunt, but also the women and girls, who would have been otherwise gainfully employed gathering or helping to drive more animals into the men's nets. And the women said they were having far too good a time in the village. So it was decided to bring the *elima* to an end.

It was during this time that the older women paid more and more attention to what was going on, and gathered almost nightly outside the house to join in the singing. But this evening was different in that almost every woman in the group was there. There were even some women from Cephu's group, which had temporarily detached itself and was now some miles away. They gathered early, before it was fully dark. The bright blue sky was deepening without any of the wild tinges of colour that so often come at dusk, and the tiny puffs of cloud that drifted across remained a shining, fluffy white, even when the sky was almost black.

The women brought sticks of fire with them, and some of them cooked and ate their evening meal right there, outside the *elima* house; others just sat around and gossiped, huddling close to the fires from habit, though in the village it was still warm. Inside the house the girls were singing, but the women outside did not join in. The men sat before their huts, watching and waiting. Finally the door opened and the girls trooped out. The younger ones came first, their bodies decorated with modest blobs of white paste made from clay; then came Akidi-nimba and Kidaya and their attendants, resplendent and magnificent, yet strangely shy. They must have spent hours working on themselves and each other, for they were covered with elaborate patterns skilfully drawn with fingers and thin sticks, also using the white clay. Akidinimba had taken par-ticular trouble to paint her breasts with intricate circles of lines punctuated with blobs, something like a grape-vine gone to seed, whereas the more demure Kidaya had concentrated on her buttocks, and these were covered with a hundred carefully painted little stars. Each girl had her favourite style, and each was different. They looked around coyly, to make sure they were being admired, and sat down at the edge of the group of women.

They began singing then, together with the women, much more seriously than they had ever done before. They sang as though they *were* women. They sang right through the evening, and now and then stood up and danced in a bashful circle around one of the fires. They sang songs whose words had no particular significance, but which in themselves were of the greatest significance, being songs only sung by adult women. That was why so many of the mothers and grandmothers had assembled, to welcome their daughters into their midst again, no longer as children, but as friends and partners in adult life. Some of the women looked sad, for they knew that this meant that the girls would soon get married, and then they would almost certainly leave to join their husbands in other camps.

Akidinimba was very different from her usual boisterous self. She was still self-assured, and she bounced her breasts up and down as flagrantly as ever; but perhaps she was subdued by the thought that she was no longer the leader of a group of young girls, but only a young and inexperienced member of womanhood, with all the responsibilities this entailed. Kidaya, too, was thoughtful. While the others sang Kidaya sat with her back to them, fondling a large clay cooking pot. She tapped on it with her fingers, keeping time to the music. Then she drew it close and held the mouth of the pot against her stomach, bending her head down to listen to the effect. It evidently pleased her, because she continued the experiment by taking hold of her breasts and forcing them into the mouth of the pot, squirming around until she was comfortable. Then she moved her body rhythmically, producing a variety of curious sounds, not only in the tone of the drumming of her hands, but by making sucking noises as the air was forced in and out. When she tired of that she put the pot between her knees, and sang quietly into the mouth of it, getting a deep, reverberating effect. This sounded so much like the *molimo* that the men started, and one of the women turned and said something sharply to Kidaya. Kidaya herself was a little startled, and put the pot away from her. She relapsed into a thoughtful mood, slowly and deliberately rubbing herself with her hands, wiping off all the beautiful decorations so laboriously painted on her body by her friends. When the singing ended and the girls retired for the night, Akidinimba was still as decorative as when she first appeared; but Kidaya was just a mass of streaked and dirty grey. And there were tears in her eyes.

During the last week the *elima* battles reached their pitch, both sides making the most of the opportunity while it lasted. It was, on the surface, a pretty happy-go-lucky affair, and what went on inside the hut was best left to the imagination. But in fact it all had a very definite purpose, and was closely supervised by the watchful eyes of Asofalinda and Mambunia. For the pygmies, the *elima* is not just a puberty rite for girls. It is a celebration of adulthood, and just as much for boys as for girls. For Akidinimba and Kidaya certain physical changes had taken place which marked them as women, but for boys there are no such self-evident changes. They have to prove their manhood. Village boys do this by entering the initiation schools, and even though pygmies enter the same schools, it is for a different purpose, quite unconnected with adulthood as they see it. For them the *elima* serves the purpose, or at least a part of it.

In the *elima* the boy has to show considerable courage to fight his way into the house, after he has been invited. In this last week the women maintain constant guard outside, and they are well supplied with all sorts of ammunition. If they want to keep any particular boy away from their daughters they are perfectly capable of doing it. And when a boy finally succeeds in breaking his way through he still has to face a possible beating from the girls inside. If he has not been invited by having been beaten beforehand, he will certainly have to go through it now.

As well as this, in order to prove himself a man, the boy has to kill 'a real animal'. That is to say, not a small one, such as a child might kill, but one of the larger antelopes, or even a buffalo, so proving that he is not only capable of feeding his own family but also able to help feed the older members of the group who can no longer fend for themselves.

Once a boy has gained access to the hut he may do one of several things. He may flirt or even sleep with the girl who invited him. The *elima* gives an opportunity for boys and girls to get to know each other intimately, and such friendships often end in marriage. On the other hand it gives two lovers a chance to discover that they are unsuited to each other before becoming betrothed officially. From what I was told, by older men and women as well as by the bachelors who took part in this *elima*, one may even have intercourse, but with certain restrictions. These restrictions must be effective in preventing conception, because nowhere have I ever heard of a girl becoming pregnant in an *elima* house.

A boy may only sleep with a girl if she consents. The tacit consent of her mother has already been given by his having been allowed to gain access to the *elima* house. If he sleeps with her then he must not leave the house, and has to stay there until the festival is over, subject to the same restrictions as the girls. But he may decide on just a mild flirtation. Or he may, and this is considered a master move, simply turn around and leave the house without having paid the slightest attention to any of the girls. Whatever happens, there are always the older girls, such as Kondabate, to keep an eye on their younger companions and see that they do not get into trouble.

The conclusion of this particular *elima* was an anti-climax. The pygmies are not a ritualistically minded people; to them the important thing about any festival is that they should openly express their emotions and accept the realities of whatever situation the festival marks. Instead of living in constant fear of the spirits of the dead, performing elaborate rituals to remove the souls of the departed as far away as possible and as quickly as possible, the pygmies sing in their memory for months on end, during the *molimo*. And so, with the *elima*, they sing to celebrate the blessing of potential motherhood conferred on the fortunate girl or girls. The month or two that the festival lasts gives them time to adapt themselves to the new situation, but they have no need for elaborate ritual acts. When the girls appeared that night and sat down with the women, and sang

with them before all the camp, that was a formal public acknow-
ledgment of their new status as women. And yet the *elima*
dragged on for several days afterwards.

Part of the reason, of course, was that if they ended with a
feast, in the tradition of the villagers, the villagers would be so
pleased that they would probably contribute most of the food.
And so it was in this case. For all other practical purposes the
elima had come to an end the night they sang together. But the
girls had another reason, they wanted to placate the villagers.
Sabani in particular had been extremely upset about the non-
orthodox conduct of this *elima*. He, and many others, felt it was
wrong for the girls to go wandering about from village to
village, sometimes sleeping overnight in camps several miles
away. This was unheard of. It was, to the villagers, as though a
colony of lepers were wandering at large. But even more they
resented the liberties the girls took with village custom. There
is a certain dress that village *elima* girls are meant to wear,
because the special leaves of which it is made have the magical
power of protection; other dress is ineffective, and invites disaster. The pygmy girls, however, found that the ineffective
leaves were by far the most attractive. As they themselves said,
'This plant makes such pretty skirts; why should we make ourselves ugly with that plant? This is a time to rejoice.' They also
liked to imitate the village boy's initiation, and this so infuriated, and worried, the village traditionalists that they
threatened to invoke the most powerful curses on all the girls.
And so the girls put on a traditional ending (village style) to
appease them.

On a Sunday morning they came out of their house as soon
as the first light of dawn flecked the sky with gold. They went
down to the banks of a forest stream, where it widened out into
a silent pool, and bathed themselves in half-hearted imitation
of an elaborate village ritual. Then they oiled their bodies and
donned oily, new bark-cloths, and returned to dance in the
village. They danced in a tight little circle while the women
cooked the food. The villagers came to watch approvingly, and
to see that the food was properly distributed, for only certain
people may eat it. They also made sure that all the leaves
which had been used as a sleeping mattress in the *elima* house

were thrown away and the house cleaned thoroughly. In this way they satisfied themselves that no harm would come to 'their' pygmies. The ancestors would be appeased by the feast, and the evil spirit that hovered around the unclean girls would be driven away.

The girls found it all rather tedious and boring, and as soon as they were able they broke off from their dancing and went wandering around the village; Akidinimba in search of Pumba, who had failed her badly in the final ceremony. The others visited from house to house, in the hopes of gifts and compliments. The older women sat around and enjoyed the food for as long as it lasted, then they broke up and went about their business as though nothing had happened.

The men, for some time, had been talking about where they would build the next hunting camp.

Chapter Eleven

THE END of the *elima* precipitated a crisis in the village. One of the many ways in which the villagers try to exert control over the pygmies is by seeing that they are married according to negro custom. The villager believes this achieves two ends. He believes that it subjects the married couple to the same supernatural authority that controls married villagers (the authority of village ancestors), and he believes that by careful manipulation and match-making it can increase and improve his own stock of pygmies.

The pygmies play along with this, but in point of fact they consider a village wedding no more than a formal betrothal at the most, and a good excuse for a mild flirtation and a great deal of feasting. But if two pygmies want to get married, and they are near the village, then they gladly undergo the village ritual, sanctifying the bond later in their own manner. Everyone is happy; the villagers think the marriage is a success because it was performed according to their customs, while the pygmy knows it is a success because it followed his as well.

The villagers naturally looked on the *elima* as a prelude to the marriage of Akidinimba and Kidaya. The pygmies, however, regarded it in no such light. They wanted to get back to the forest, where the girls could get married in their own time and in their own way. This would involve no more than an exchange of visits between the families concerned, and the presentation by the groom to his in-laws of a large antelope, and perhaps some arrows and a bow. But none the less they were tempted by the idea of more feasting. So there was some considerable conflict of ideas and ideals, particularly as the villagers had their own ideas as to whom the girls should marry.

The *elima* had created a crisis in another direction also. Pumba, whom Akidinimba had loved so ardently and beaten so thoroughly, to show her affection, had kept his part of the bargain by coming to the *elima* house during the final week, fighting his way in as a man. But once inside, instead of making love to Akidinimba he paid court to Kidaya. Fortunately for

Kidaya, but unfortunately for Pumba, she did not respond, and without her permission Pumba had no way of forcing his attentions further. So he stalked out of the *elima* house, pursued by shrieks of rage from Akidinimba. He took his friend Kelemoke's hand briefly, then left without a word. He crossed the Epulu and went back to his camp, and he was seen no more.

Everyone was upset because they all felt that the sooner Akidinimba got married the better, and Pumba was popular. So much so that some pressure was brought to bear on Kidaya to force her to accept him, but Kidaya had set her heart on a handsome youth attached to the village of Katala, several miles beyond Eboyo. His name was Atete. So she refused to have anything to do with Pumba.

But the prime objective in everyone's mind was to get Akidinimba married, so once it was plain that Kidaya would not take Pumba, they lost interest in her and returned to the problem of her fellow initiate. Circumstance forced the issue, for while all this was going on a close relative of Akidinimba's, a boy named Mulanga, decided to get married to Atete's sister. Nobody saw any harm in this, and so arrangements were pursued and they were duly married. Mulanga was quite nonplussed by the ritual, for he had spent most of his life in the forest and knew little of village customs. He was particularly perturbed when at the end of the ceremony his bride was seized by a village woman and led away. He was not quite sure what to do, and trotted behind them at a safe distance until he was able to follow his wife into the house that had been set aside for him.

Mulanga's marriage meant that he was obliged to provide a sister for his bride's brother, Atete, to marry. Villagers will compensate families for the loss of a daughter by the payment of money or goods, but the pygmies find it necessary to arrange for an exchange, so that no group is ever depleted of its womenfolk. Akidinimba, who had still not forgotten how she had been jilted and insulted, saw in this an opportunity for revenging herself on Kidaya, the unwitting receiver of Pumba's attentions. So she volunteered to marry Atete. Everyone, except Kidaya, was delighted, and Atete and his family agreed. By the time the date for the wedding was set, Akidinimba began to have doubts, and thinking of the freedom she was about to lose she began to weep day and night. But there was no escape, for the villagers of Katala had arranged a great feast and this made everyone all the more anxious to marry her off.

Atete, however, heard of some honey way off in the forest, several days' march away, and at the last moment he set off. The villagers protested loudly, but Atete merely said that the wedding could take place any time, but the honey could not wait. As it was only a pygmy wedding the villagers shrugged their shoulders and forgot to notify Akidinimba. So on the appointed day, when the young bride-to-be had been painted and richly oiled, she set off, weeping hard, to find when she arrived that there was no sign of Atete, no indication of any of the promised preparations for a wedding, and nobody who had the least idea when Atete might be back. Akidinimba brightened up at once, and suggested returning to Epulu. But having got that far, her relatives decided to stay. Besides, this was Ngoma's village and they could be sure of being fed well.

An elderly village woman swept out a dark and dirty two-roomed house in the middle of the village, and told the bridal party that it could wait there. She reluctantly went to see that food was cooked for them, and word was sent out to Atete, in the forest, to return quickly before the bridal party ate up the entire village. By nightfall there was no word, and everyone decided that they were being fed so well that they would stay over and see what happened the next day. They sang for a while, but things had fallen rather flat and it did not last long. When the others went to sleep Akidinimba was crying once again.

The next day dragged its hot and dusty way along, and still there was no sign of the groom. The party began to get as restive as the villagers, but still they decided to stay. Mulanga arrived to try and persuade his young wife to return with him, but she refused, saying she was not going to miss her brother's wedding for any husband. He retorted that it looked as though there was not going to be any wedding, to which she replied, 'In which case you will have no bride.' This is one of the difficulties of sister-exchange marriage, for if one marriage breaks up and the woman returns to her home, then the corresponding marriage also breaks up, the injured party claiming back its sister. Mulanga, alarmed at the prospect, set about beating his sister, asking the unfortunate Akidinimba why Atete did not come, as though it were her fault. He even went so far as to say that it was no wonder nobody wanted to marry her, the way she flirted. Under other circumstances Akidinimba would have attacked her brother, for she was twice his size; but she had lost all her bounce and was so miserable by this time that she just burst into tears once more, rolling over and over on the dirty floor. The second day ended more unhappily than the first, with still no news from Atete. The messenger that had been sent out had himself not returned.

Early next morning, however, the messenger came back with several leaf bundles full of honey, and announced that Atete was on his way, and would be arriving soon. Villagers and pygmies alike began making preparations feverishly, but by the time Atete arrived it was too late to do more than have a dance, and the actual wedding was postponed for yet another day. By this time the villagers were thoroughly tired of having to feed a large wedding party for so long, and they went through the ritual just as fast as they could. The wedding began in the morning, and Akidinimba cheered up immensely when some-one pointed out to her that it is better to be married than to be jilted, particularly when you are all dressed for the wedding. She laughed and giggled all the time that an old village woman was telling her, as custom demands, how to behave towards her husband, what a serious business this was, and how she would be struck dead if she did not adhere to the rules, so that finally the old woman gave up and just pushed her into the bridal house, where Atete was waiting, and slammed the door on them. Shortly afterwards a rather subdued Akidinimba emerged, and began cooking her husband's supper.

There was a great deal of speculation amongst the pygmies as to whether or not this union would last, but it was settled when, one day, Atete arrived and presented his in-laws with a fine *sondu* antelope. He said that he was very content with his wife, and he had not found it necessary to beat her at all often.

* * *

Sometimes pygmies talked about the system of bride-wealth used by the villagers, and sometimes they wished that they could use the same system. But for the villager the woman is a labourer in the fields who can cultivate crops that can be sold for cash. From the village groom's point of view it is important to make a satisfactory payment as it not only enhances his prestige, but it also gives him the assurance that his in-laws will be on his side, not wanting to have to return so much wealth should their daughter change her mind. But for the pygmies this is unworkable. To them a woman is more than a mere producer of wealth. She is an essential partner in their economy. Without a wife a man cannot hunt, he has no hearth, he has nobody to build his house, gather fruits and vegetables and cook for him. So a group that loses a woman looks for a woman in return. The pygmies extend the term 'brother' or 'sister' so that it can be applied to almost anyone of the same age group, however distantly related, and this means it is seldom impossible to find someone to exchange, and someone who is willing. But sometimes it needs a little coercion. So it was with my friend Kenge.

Kenge often laughed at the villagers, and the way they sponsor a pygmy wedding, even paying bride-wealth to each other—the 'owner' of the groom to the 'owner' of the bride. He found it particularly funny when later the marriage broke up, because the two pygmies considered it a mere flirtation, and the villagers then had to fight over the necessary restitution of wealth. But he wished he had been able to use the system to get himself a wife.

He was deeply in love with a particularly attractive young girl named Maliamo, and he badly wanted to marry her. He had three sisters. But one of them was married to Masimongo, one was still too young, and the other was unwilling. Yambabo, the unwilling one, was a very determined girl, as stubborn as Kenge himself. She was quite fond of Maliamo's brother, a slim, light-skinned youth called Taphu. But nothing in the world would make her admit it once she saw that Kenge wanted to use her in order to get Maliamo. Kenge's decision came almost as much as a surprise to her as it did to the rest of us, as he was as notorious as Akidinimba for being a confirmed flirt. In fact most people thought he would marry Biyo, Tungana's daughter. But Kenge said no, Tungana was almost like a relative to him, and it would be incest to marry his daughter. People pointed out that this had not stopped him from sleeping with Biyo, but Kenge maintained that the two situations were totally different. Anyway, he had set his heart on Maliamo, and she had set her heart on him.

The situation dragged on for so long that Kenge began to see rather more of Maliamo than was allowed by customary flirtation, and people started to whisper. This drove him to take more desperate measures with his sister.

One morning the village camp was awakened to the sound of terrified screaming from the house directly opposite mine, where Yambabo was sleeping. I looked out of my window and saw Kenge dragging his sister out of the hut by one arm, pulling her over the ground and shouting to the camp that she was no good and should be killed. He pointed to her breasts and said that she had enough milk to feed a dozen children, why did she refuse to marry? Yambabo was as strong as a buffalo, he continued, so why did she refuse to work? He then gave what he considered could be the only reason, which was extremely personal and uncomplimentary. Yambabo tried to get to her feet to hit him, but every time she began struggling he simply thumped her on the back with his fist, still keeping a tight hold of her with his free hand.

People came sleepily out of their huts to watch, all rather agreeing that Yambabo really should have married long ago, and deserved a brotherly beating. Encouraged by this, Kenge began kicking her, and she responded by biting him in the leg. Moke tried to intervene, but it was no good. Kenge was ready for murder, and by the time he had finished with Yambabo she was a sorry sight, scratched and bleeding, with one eye swollen. And still she refused to marry Taphu.

From that morning onwards we all came to accept that Kenge was going to continue beating his sister until she gave in and it was just a question of how long she could hold out. One day, when her brother had been particularly bad-tempered with her she ran away to their mother's home, many miles to the east. I asked Kenge if he was going to follow her, but he shook his head. He said, 'Our mother will beat her much more severely than I, for she knows that Yambabo loves Taphu, but is too lazy to get married. All she wants to do is to sleep with him. She is as bad as Akidinimba.' And so he stayed, sitting by the fire all day in a sulk, leaving me to cook my own food and his.

That evening the old men discussed the case by the central fire, with the women throwing in remarks from the huts all around. They were unanimous in believing that Kenge had done the only thing he could have done. Perhaps he should have beaten her harder, Tungana said, for some girls like being beaten. And after all, it was not as though he had chosen her husband, as the villagers do, sometimes forcing a young girl to marry an old and ugly man merely because he can pay more bride-wealth than a younger man. Everyone knew that Taphu loved Yambabo as much as she loved him. He had asked her to marry him half a dozen times at least, and was constantly giving little presents of meat and honey to her mother.

The next morning, shortly after dawn, I heard loud weeping and saw a bedraggled Yambabo being marched into the camp by an irate mother, Badorinji. Kenge was wandering about searching for firewood, but he hardly even looked up. He just muttered: 'Now you will see, I did not beat her for myself alone.'

Badorinji led her daughter by the wrist into the middle of the little village. She called out in her loud, musical voice, that her son had always been a good son, and had been a father to his sisters ever since their own father had died, and had been a man to her. And now that he wanted to have a wife of his own she had made him a hunting net, but what did this worthless girl do? She gave Yambabo a hefty slap. Yambabo, however, had lost all her fight, and could only whimper. When she had reached her mother's village she had promptly been turned around and made to walk all the way back, right through the night. Badorinji continued, telling everyone that her daughter had been a no-good right from the start; all she liked doing was eating, sleeping, and going with boys. She would do no work, and she had refused Taphu, who loved her, many, many times. Then Badorinji ended with a masterpiece of statesmanship. She said that she would betroth her youngest daughter to Taphu, which would allow Kenge to marry Maliamo, and that she would have nothing further to do with Yambabo. Yambabo would lose Taphu, and without her mother's help she would never be able to make a respectable marriage, for she had no other brothers. She could find some-one to marry her, but unless a girl were given in exchange for her she would have no status and no security. Only her mother or Kenge could give her away and receive a girl in exchange.

Yambabo let out a cry of protest at this, and said that she was tired and wanted to lie down, why did everyone shout at her. Her mother slapped her once more and asked her if she would marry Taphu or not. Yambabo, with a final wail of protest, swearing that everyone had treated her so badly that she would surely die, gave up the battle and said that she would marry Taphu, or anyone else for that matter.

For the first time Kenge stopped trying to pretend that he was not listening to what was going on. He called a young boy and told him to run to Babama village with the news, and to tell them to get ready to receive Yambabo's wedding party. He added, hopefully, 'And tell Maliamo that she is my wife now and I want her to come here at once and cook my meals for me. I am too busy to go and collect her.' He then went up to his mother and greeted her, which he had not done since she entered the camp. She put her arms around him, and it was strange to see the proud, sophisticated and very adult Kenge allowing himself to be embraced with such a show of affection. But what Badorinji had said was true, he was a good son. And he was a good brother too, for when Badorinji had finished hugging him, doing a little dance at the same time, he went and took the weeping Yambabo's hand and led her to the hut oppo-site mine, and there he put his arm around her and whispered something in her ear as he sent her inside to lie down and sleep.

Taphu sent word back that Yambabo would have to hurry as he was off to the forest and would not be back for a long time. Maliamo also sent a message, to say that the villagers would not let her come until Yambabo had married Taphu, because they did not trust Yambabo. The villagers announced their opinion

that Kenge had already slept so much with Maliamo that it
would not hurt him to wait. Kenge was furious, because this
meant that he would have to go through a village wedding,
unless he wanted to alienate the villagers by ignoring them. He
could have done this, but that would have meant losing good
exchange-friends there who kept him supplied with village
goods in return for a few favours in the forest. Kenge enjoyed
all the fun and games of a village wedding when someone else
was involved, and he was content for Yambabo to go through
the ceremony, but for himself he wanted no part of stupid
village customs. He was going to carry Maliamo away and
give her parents an antelope, or maybe an elephant, like any
self-respecting pygmy. He began dreaming out aloud, describ-
ing exactly how he was going to track the elephant, and maybe
kill a buffalo or two on the way, and he only came back to
earth when I asked him how he was going to bring the elephant back to the village. He said that I would see, he was
going to get Maliamo right now. He set off at once. The next
day he returned, alone. Maliamo had gone into hiding, not
wanting to jeopardize her brother's chances by going to live
with Kenge before Yambabo had been well and truly bound
to Taphu.

There was only one thing to do, so Kenge made arrangements for Yambabo to leave the next day, with a suitable
wedding party. They would stop overnight at Katala's village,
and continue the next day to Babama. At this point an uncle
of Taphu died, and the Babama villagers sent a message to
say that the wedding was off. Taphu, however, sent word to
say that as far as the pygmies were concerned the death made
no difference, and the wedding was on. Kenge did not want to
offend the villagers at this stage, so he delayed a few days, and
after a respectable lapse of time he finally sent the wedding
party off. But not before Yambabo had had yet another change
of mind. As she was being rubbed with palm oil and camwood
paste, so that she would look her very best, she suddenly
reached out for the hearth. Before anyone could stop her she
had covered herself from head to foot in ashes, the sign of
deepest mourning. As the rest of the party had already been
oiled and painted for some time, and were anxious to be on

their way, there was not much sympathy for Yambabo. They told her that if she wanted to walk through the villages like a ghost they did not mind, but one way or another she was going. Masisi, who was acting as her father, gave her a long lecture during which his wife hurriedly wiped off the ashes and smeared on more oil.

The party finally got off and had a slow and pleasant journey, stopping in each village in the hopes of receiving presents of food and palm wine. It stopped over at Katala's, where the long-suffering family of Ngoma fed them all lavishly, and the next day it arrived at Babama. Yambabo was led through the village to where the pygmies had their encampment, but the villagers came hurrying after, saying the wedding could not possibly take place there so soon after the death of a relative. It would have to be held in the village. Taphu wandered out of his hut to see what was going on. He was dressed in a dreadful array of cast-off clothes given him by the villagers. He did not seem particularly interested, because after listening to the argument for half a minute he yawned and went back into his hut without saying a word to anyone.

We all trekked back to the village, and there, under the watchful eye of the villagers, the wedding began. Masisi had refused to come, saying the Babama villagers were the worst savages of all, so Mambunia came in his place. He supervised as all the wedding gifts were placed on the ground, and he watched as various villagers, each in the hopes of improving his chances of getting some meat from the young couple, came forward and added either money, axe-blades, or small pieces of cloth. There were already several pots, quite a number of arrows, and a few trinkets given to the party at Katala village. The villagers who claimed Maliamo felt that this treasure belonged to them, but Mambunia denied this strongly, and pointed out that in any case, according to their own village custom, the gifts should sleep with the couple that night. Reluctantly the villagers agreed, and Mambunia promptly scooped everything up and brought it into the wedding house.

Taphu was meant to put in an appearance at this point, but he could not be found. It was two hours before he was discovered, asleep, and dragged to where we were all sitting around, eating and drinking and singing. He arrived in his tattered village clothes, but was promptly taken inside and made to change into a new cloth, put on a woven straw hat decorated with feathers, and arm himself with a new bow and a sheaf of arrows. Arrayed like this he appeared at the doorway in the late afternoon. Yambabo looked up and actually smiled. Taphu looked very handsome and imposing, if somewhat uninterested, and having made his duty appearance he went back inside and fell asleep again.

That night the wedding party stayed at Babama, all sleeping in the wedding house, feasted royally by the villagers. The next day the rites were completed in a ceremony which consisted largely of the formal appearance of Yambabo, standing between two of her sisters, preceded by a still disinterested looking groom. As one villager after another lectured the bride, Taphu shifted his weight listlessly from one foot to the other, scratched his back with his new arrows, and gazed longingly at some monkeys in the trees nearby. The villagers warned Yambabo that she belonged to them now, and could not go and call on Kenge whenever she wanted help, she would have to come to them. They pointed out which villagers she was expected to help when she and her husband came from the forest, and she was told of all the things they would expect her to do for them. She was finally reminded of the villagers that had contributed gifts for the wedding.

Yambabo and Taphu were led back into the house, and the others resumed their singing and dancing and drinking. Munyunga, the burly villager who had done most of the talking, and who claimed to 'own' Taphu and Maliamo, said that he wanted to wait a while, according to custom, and then bring Maliamo with a large wedding party to Kenge's village. But to this the pygmies as one man absolutely refused to agree. There was a great deal of argument, and it was plain that Munyunga was determined to have, at least, a return feast arranged for him. Normally he would have arranged this with Kenge's 'owner', but this was wily old Kapapela, the Paramount Chief's brother, and he had already made it known to Kenge that he was in complete agreement with Kenge's refusal to have a village wedding, thus saving himself the expense. So

Munyunga knew his chances were slim, particularly as Kenge was no longer even at Kapapela's village and refused to go there. So he finally agreed to arrange matters with Kapapela later, and in the meantime he would come to Kenge and bring Maliamo with a small party, but that he would expect food, drink, and some kind of gift. He suggested a new hat. He agreed to come the next day, so we returned to Camp Putnam well pleased.

Munyunga arrived early the following morning, with Maliamo and an escort of three villagers. He was clearly annoyed by the fact that only Moke and Masisi were there to greet him, everyone else having disappeared, many of them to hunt and others simply because they did not like Munyunga. He was even more upset when Masisi, instead of greeting him first, and then the villagers, went straight to Maliamo and led her away to his hut so that she could put down her bundle and refresh herself. Kenge was nowhere to be found. Moke invited the villagers to sit on my *baraza*, and they made loud remarks amongst themselves that as I was Kenge's new 'owner' I should offer them beer. But I wanted to see what Kenge would do on his own, so although I had several containers of palm wine waiting inside the hut I did nothing.

Kenge was finally found, engrossed in a gambling game behind Ekianga's house. He had hoped that the villagers would become discouraged and leave. But after two hours of patient waiting he emerged and slowly wandered across the clearing towards us. He stopped to look inside several huts on the way, chatting to anyone he could find, as though he was completely unaware that the wedding party was waiting for him. He was as slow and as surly as he could be, and when he finally reached the villagers, who by then were not even trying to disguise their anger, he just took each of their hands in turn and then set his back to them and asked me what I would like to have for lunch.

This was too much for Munyunga, who stood up and shouted that he was going to take Maliamo back with him, that Kenge was a child who did not know how to behave like an adult. This touched Kenge on a very sore spot, for he prided himself on being extremely adult. He motioned Munyunga to sit down again, and said, 'You are only villagers—Ebamunyma is a white man. Isn't it right that I should ask him first?' I hastily said that I could wait, and without another word Kenge went away. About half an hour later he returned with two bottles of beer, some dried fish, and a bottle of palm oil. He put these on the ground beside Munyunga, and then wandered out into the middle of the village, where he played with the children. The next thing I knew he had disappeared again, and I was told he had stolen one of the villager's bicycles and gone off down the road to Katala.

The pygmies were obviously going to have nothing further to do with the villagers, and I felt sorry for Maliamo who had come and sat down beside them in the *baraza*. Feeling that I was in some way responsible, I asked them if it was proper for me to offer them some palm wine, and this eased the situation slightly. Munyunga and his friends stayed and drank their beer and palm wine, while Masisi and Moke did their best to pacify them by offering to have a meal cooked. But they said no, all they wanted was to drink. Only two pygmies came to join them, and then only to see if they would be offered any leftovers. But Munyunga was not parting with a drop, even for Maliamo, who continued to sit uncomplainingly by his side. When the beer and palm wine were all finished, Munyunga still stayed; angry, surly, and very drunk indeed. Occasionally he called for more to drink, or asked where his new hat was, but nobody answered him. The few people that were in the village just went their way as though there were no visitors at all. Moke was fast asleep outside his hut, and Masisi was fashioning a new bow, never as much as glancing at the *baraza*.

It was beginning to darken when the villagers left, saying that the pygmies were a lot of animals who didn't know how to behave, and that if Maliamo did not bear Kenge any children it would not be their fault.

As soon as the unwelcome guests had left, the whole atmosphere changed. Pygmies appeared unexpectedly from all directions, crowding around Maliamo to greet her. She went to the little hut that had been given for her and Kenge to live in, and began cooking. Kenge appeared, as if by magic, laughing and telling how he had met Munyunga on the road, and that at the rate he and his friends were going it would take them three days to get back to Babama.

When evening fell, Kenge went up to Maliamo for the first time during the entire day, and sat down with her as she dished out the food she had cooked. He brought me over a bowl, and said he wanted me to try his wife's cooking. He sat beside me and ate, as he often did, from the same dish. I was a little curious that he had not paid more attention to Maliamo and I asked him about it. He laughed and replied that Maliamo was no stranger to him; the only difference now was that following Yambabo's marriage to Taphu, Maliamo was able to come to his village instead of his having to go to hers. He had given her father an antelope, why did the villagers have to make all this fuss? He abruptly returned the conversation to more important things. Did I or did I not like his wife's cooking? I did.

At the central fire everyone was discussing the way Kenge had treated the villagers. They were mostly secretly delighted, but Masisi said that even if Munyunga *was* a drunken fool, there was no need to be quite so rude to him; that Kenge should have stayed, as he and Moke had done, and merely sat and said nothing. Mulanga, remembering his own wedding, said that he could see no sense in a village wedding anyway, but Masisi pointed out, 'If we did not let the villagers think that they arrange our marriages, then how could we expect them to give us such fine feasts and present us with so many gifts?' Moke, picking up his pipe and passing it to his son to be filled, added, 'After all, it doesn't really make any difference; you are not married properly until you give your in-laws a forest antelope. If you don't like your village wedding, don't give the antelope.'

And so, everyone agreed, a village wedding was great fun, but '*bule*'—empty. It was completely empty; completely and absolutely. With that they laughed loudly and slapped each other on the back, and forgot all about it.

Chapter Twelve

OF ALL the ways in which the villagers attempt to assert some kind of control over the BaMbuti, the one which is the most significant, and which they consider to be the most effective and binding, is that of subjecting pygmy boys to the village initiation, known as the *nkumbi*. And this is one custom which, at first glance, does not seem to have any counterpart in the forest life of the pygmies. Most outside observers have assumed that the very fact of participation by the pygmies in the *nkumbi*, and the absence of any such initiation rites of their own, is a sign of their dependence on the village. Others go so far as to say that it only shows they are a people devoid of any culture of their own. If the facts were what they seemed to be these observations might have held water; but the real facts are very different.

I had the good fortune to stumble over fresh information, quite accidentally, which almost completely reversed the picture. That was during a previous visit to Epulu, in 1954, which I mentioned briefly earlier (Chapter One). Three years had elapsed since the previous *nkumbi*, in 1951, and it was time for another. The purpose of the ceremony is not only to fit village boys for adult life, making them full members of the tribe, but also to ensure the continuity between the past, present and future life of the tribe; between the living and the dead. To the villagers the dead, the ancestors, are just as important as the living, and in many ways more so; and it is only those who are initiated that will join the ancestors when they die. Just as it is important to have a thriving, populous tribe of the living, so is it important to have a thriving, populous tribe of the dead. The dead are, in fact, living—but in another world. Regular initiations are essential to ensure the continued growth of that far-off tribe of ancestors. I have even seen a young boy who had been due to enter the *nkumbi*, but who died just before it took place, circumcised and accorded full initiation rites before burial, just as though he were alive. In this way he was able to go to join his fathers.

In 1954 there was a unique situation. There was only one village boy of the right age for initiation (usually between the ages of eight and twelve in this area), and his father was a Christian who refused to allow the boy to enter. Every pressure was brought on the family, but to no avail. There were eight pygmy boys of suitable age, and normally there would have been at least as many villagers. They would all have been initiated together, but in such a way as to make clear the inferior status of the pygmy. One of the initial acts is the circumcision of the boys, and a pygmy boy is sent first 'to clean the knife'. Then follows a villager, and these two are bound together by an indissoluble bond, the bond of *kare* blood-brotherhood. They are mutually obliged to help and care for each other for the rest of their lives. This, of course, helps to reinforce practical and economic links between the two people, as a pygmy theoretically would be unable to refuse his *kare* meat, and the *kare* could not refuse any demand for plantains, manioc or rice.

But this year the villagers were faced with the alternative of either having no initiation at all, or having one for pygmy boys alone. As the initiation of a pygmy only gives him partial rights, not being a full member of the tribe, a pygmy initiate can never take a leading part in the *nkumbi*; all the cutting, and all the formal teaching and certainly all the ritual acts can only be performed by a qualified villager. So it meant that the villagers would have to go through all the work and expense purely for the pygmies. There could be no question of not holding the *nkumbi*, as this would offend the ancestors. And in any case the villagers decided it was worth their while as it helped to assert their authority over the pygmies. I heard speculative talk more than once about how an initiated pygmy, when he dies, goes to serve the tribal ancestors just as he served the villagers during his life.

The *makata* sticks were cut and brought out. These are a series of sticks used only at the time of the *nkumbi*; each one is cut to a different length and tuned to a definite pitch. Some eight or nine men, each with a stick under his left arm, tapping it with a shorter stick held in the right hand, dance around the village playing a complicated but tuneful rhythm, the different notes overlapping and combining. Villagers from many miles around, some of them from as far as fifty miles away, congregated at Epulu, and dancing and feasting went on for over a week. During that time the boys were collected together, and last-minute appeals were made for young Yusefu, the village boy, to be allowed to join. His father settled the matter by taking him away and sending him to a mission school the far side of Mambasa.

The boys were not all cut on the same day, but over a period of two weeks. Akobo, the youngest son of Masisi, was the first to be cut, and with him went Massoudi, nephew of Ekianga. They were, at a tender ten or eleven years of age, as brave as any adult, but they were particularly tough children. It was different with little Sansiwake, one of Cephu's nephews. Sansiwake was one of the most beautiful children imaginable, and as beautiful and gentle in character as he was in appearance. He was soft-spoken, and had a voice as musical as his name. His eyes were the dark, melting eyes of the *boloko* antelope, as sensitive and as trusting. (On this occasion they were as frightened as those of any young animal caught in a hunting net, struggling for its life.) His body was as graceful and as delicate as that of a fawn, and his thoughts were only thoughts of kindness, of concern and love for his mother, Tamasa, whose face was so eaten away by yaws; and for his father, Aberi, who was so ugly and who constantly brought ridicule on himself and his family.

Sansiwake was the only boy to cry when being taken away from his home. He cried not only for himself, but because his mother was crying bitterly at being parted from her son who was as much a man to her as her husband. Sansiwake was brought to the village and his head was shaved as the first of a series of ritual acts symbolizing the casting off of an old life. Then Sabani appeared, dressed and painted in a terrifying manner, to smell out his next victim. His face was covered with a mask, his body was daubed with white clay, and his arms and legs were bound with great bundles of raffia so that as he danced he looked half-animal and half-bird. With him were a number of *nkumbi* drummers, beating a wild and ear-shattering tattoo, driving all women and uninitiated children into their houses, where they locked and barred the doors. Only little Sansiwake was left outside.

After Sabani had danced around him, with the drums getting louder and louder, he was picked up and carried by a villager to the stream where the cutting was to take place. Some boys were cut in the initiation camp, others in the centre of the village, and some in the stream; it depended largely on the whim of the moment. In the camp there was a scaffold on which the boys sat while the doctor cut, but down in the stream, a short distance into the forest, Sansiwake had to stand on the rocks, the cold water splashing over his feet. He was held by the arms and legs so that he would not be able to struggle, and a considerate uncle, Masalito, clapped a hand over his nephew's mouth so that his screams would not bring him disgrace in the eyes of the villagers. But Sansiwake could not stand the pain. Perhaps Sabani was drunk and cut badly; but Sansiwake bit Masalito and then screamed his head off, struggling until he almost broke loose. And two days later when he was just beginning to recover, it was decided that he would have to be cut again as Sabani had not done as he should. Shortly afterwards the weakened and thoroughly dejected boy went down with pneumonia, and would have died if the pygmies had not broken all the rules of the initiation camp and kept him warm and well fed.

He was the only one to suffer any ill effects, and his behaviour would have brought him lasting disgrace had he been a villager. The pygmies, however, merely reacted by showing consideration and sympathy for Sansiwake, and by criticizing the villagers openly for their unnecessary carelessness and brutality. The villagers retorted that Sansiwake's illness was only the penalty for his not having behaved himself properly.

From the moment the boys enter the camp they are meant to stand on their own feet without any help from their relatives, but the older brothers and fathers of these boys were constantly comforting and encouraging them, at the risk of being ridiculed themselves. Masalito, in particular, was not afraid of being sneered at by the villagers. One boy, Kaoya, who was cut in the camp, was beaten as he made his way painfully to join the other boys, all sitting in a line on logs. Immediately after being cut they are made to sit down and join the others in singing one of the many work songs they will have to learn during the coming months. This takes their mind off the pain, which is all the more considerable because after the operation the wound is wrapped in a leaf containing native medicine that is full of salt. Kaoya was very roughly handled, and Masalito immediately ran up to him and put his arms around him, hugging him tight to give him strength. The villagers howled with laughter, but Masalito was more concerned with Kaoya than with them.

Not long afterwards the boys are made to dance, and then put to rest on a special bed made of split logs, laid side by side on a rough frame that keeps them two feet or so off the ground. The boys lie face downwards along the length of the logs, their heads resting on pillows made of plantain leaves. Over the bed there was a rough shelter, sloping down to the ground behind, but open all along the front and on two sides. From the roof was hung a sacred banana, which initiates and instructors would set swinging as an understood command for the boys to start singing. The boys were forbidden to touch the banana, being told they would die if they did.

There were all sorts of other taboos, or restrictions. One was that to get wet, particularly by being outside when it rained, was sure to bring to death. To eat certain foods was strictly prohibited. The boys had to eat without using their hands directly, but by spearing the food with sticks. They were forbidden to eat with their initiated relatives. The sound of the bull-roarer, a piece of wood that makes a strange, whirring noise when spun around on the end of a cord, was meant to be the voice of a forest demon, and the boys had to show respect and terror whenever they heard it.

During the day time, when the villagers were present, either as instructors or just as visitors, the boys adhered faithfully to all these and many other rules.

By their own rules, however, the villagers were not allowed to sleep in the camp at night. Only the 'fathers', that being understood to mean the real fathers and the older brothers of the boys, were allowed to sleep in the 'place of the *nkumbi*'. That meant, in this case, only pygmies. The BaMbuti had different notions, and recognizing that I was anxious to be with them right through the whole period of seclusion, they invited me to join them. After all, they said, I was really the father of all the eight boys. In this way I became one of the *baganza*, the people of the initiation.

It was not done deliberately to slight the villagers; the pygmies merely thought it would be fun, particularly as they were sure that I would keep them supplied with tobacco and palm wine and other luxuries. From their point of view no harm was being done. From the villager's point of view, how-ever, this breach of custom was likely to bring disaster not only on myself but also on the rest of the *baganza*. Beyond issuing the gravest warnings they made no other attempt to discourage me, and soon came to accept my presence. So from the first day to the last I was with the *nkumbi*, day and night.

It was in this way that at night time, when all the villagers had left, I learned what the *nkumbi* really means to the people of the forest. The moment their instructors were gone, and only pygmies were left, the boys leapt from the bed where they were meant to be sleeping, and joined their fathers around the fire, eating forbidden foods in forbidden company and in a forbid-den manner, using their fingers. One boy jumped on to a log and swung his arms around in imitation of someone swinging a bull-roarer, which of course he was meant never to have seen. Others played their favourite game of punchball with the sacred banana. A shower of rain was an invitation to run out-side and wash off all the dirt that had accumulated during the day. The villagers not only forbade the candidates to get wet, even by washing, but they smeared them constantly with white clay, from head to foot, as a sign of their death as child-ren. To the naturally clean pygmy this was intolerable, and the boys were encouraged by their fathers to take the first oppor-tunity to wash it off. When the villagers questioned them in the morning, the boys invariably replied that they had been so cold during the night that they had huddled close together, and the clay had been rubbed off in this way. In fact the men pro-vided blankets for the boys, keeping themselves warm by lying close to the fires. Kenge and I slept at one end of the shelter, half underneath the log bed; most of the other adults slept outside.

Both the boys and their fathers enjoyed the chance to make fun, in a friendly way, of the villagers, but that was not their sole reason for deliberately breaking all the taboos. They behaved as they did because to them the restrictions were not only meaningless, but belonged to a hostile world. The villagers hoped that the *nkumbi* would place the pygmies directly under the supernatural authority of the village tribal ancestors, the pygmies naturally took good care that nothing of the sort happened, proving it to themselves by this conscious flaunting of custom.

The villager believed that the initiate, pygmy or otherwise, is everlastingly bound thereafter by all the laws of the tribe, sacred and secular. He is put into direct relationship with the supernatural, whose representatives on earth are the villagers themselves. If any pygmy initiate offends a villager, then he is also offending the supernatural, the ancestors, and will be duly punished by them. The villagers live in such fear of the supernatural, with its power to bring down on offenders the curses of leprosy, yaws, dysentery and other diseases, or to cause the offender to be injured by a falling tree, that they cannot conceive of any initiate daring to offend the ancestors. That is why they believe that through the *nkumbi* they gain ultimate control over their ever-troublesome servants who refuse to behave as servants should.

From the pygmy standpoint it is not quite so obvious why they subject themselves to what they plainly consider an unnecessarily brutal ritual. The trials of the *nkumbi* candidate only begin with the actual circumcision. During the succeeding months, now usually only two or three, instead of the former six or even twelve months, the boys are subjected to one form of mild torture after another. The torture is sometimes mental rather than physical. For instance, one boy whose father refused point-blank to have anything to do with what he called, 'an empty, savage custom', but allowed his son to enter, was made to ridicule his father almost every daylight hour. The gentle little Sansiwake, who insisted on getting up from the hard, wooden bed to rejoin his companions long before he was completely well, was made to pretend that he was big and strong, carrying heavy loads and imitating the performance of acts degrading to any pygmy, but particularly so to this sensitive child. Each boy was put through some kind of torture of this nature.

The physical ordeals sometimes start out as games, but develop into cruel tests of physical endurance. A crouching dance that might be fun for a few minutes becomes agony after half an hour. A mild switching on the underside of the arm with light sticks is of no concern until, after several days, the skin becomes raw. And then the villagers notch the sticks so that they fold over and pinch the skin sharply, often drawing blood. When the boys have become used to being beaten with leafy branches, thorny bushes are substituted.

To the pygmy this all seems harsh and unnecessary, and as far as their own children are concerned they keep a strict watch over them to see that the villagers do not go to the length that they sometimes do with village children, even if this brings them into some contempt. But to the villager this toughening-up process is essential, and does not come naturally in the course of village life. The child has to be fitted for adult life, and this is what the *nkumbi* sets out to achieve. In a few months a boy becomes a man, tough and strong, physically and mentally. The process is not a pleasant one, but it is the only way in which, under tribal conditions, the goal can be achieved. The *nkumbi* achieves still other things for the villagers, and this much at least the pygmy can understand and appreciate. But in his own life the very nature of his nomadic hunting and gathering existence provides, naturally, all the toughening up and education that is needed. Children begin climbing trees sometimes before they can walk. Their muscles develop, and they overcome fear in a number of daring tree-games. Adult activities are learned from an early age by observation and initiation, for the pygmies live an open life. Their life is as open inside their tiny one-room leaf huts as it is in the middle of a forest clearing, and so the children have no need of the sex instruction which forms so large a part of the teaching given to village boys during the *nkumbi*.

The pygmies themselves say that the toughening-up effect of the *nkumbi* is salutary, though they openly deplore and frequently reject its severity. And when this particular *nkumbi* was over, the change in every one of the eight boys was plain to see. But this is not their major reason for allowing themselves to be subjected to village initiation; nor is there any question of their being subjected by force. They enter the *nkumbi* voluntarily because they are a proud people. They live necessarily in contact with the villagers, and spend a number of weeks every year either in the village or amongst villagers. These people look down on any adult who is not initiated as though he were a child; he is treated with no more respect, accorded no more privileges, and he can not take part in any adult activities. By entering the *nkumbi* the pygmies prove their adulthood to the villagers and become accepted as such and this is the only way they can. When the initiation was over even Sansiwake had the right to enter any adult *baraza* and take part in the proceedings. This is what the pygmies want from the *nkumbi*, and what they get . . . the status of adults in village eyes.

There are many things in their behaviour to show that the *nkumbi* means no more than this to them, quite apart from their deliberate lack of respect for the laws of the *nkumbi* and their irreverence for everything that is sacred to the villager. When this particular *nkumbi* was over and everyone returned to the forest, the same boys who had been walking freely amongst the men of the village, as men, went straight back to their mothers. They became children once more, and were accorded none of the privileges of the adult forest pygmy. Above all they were not allowed to take part in the singing of the men's *molimo* songs. They may have become adults in the village, but in the forest they were still children.

Far from illustrating the dependence of the pygmies upon the villagers, the *nkumbi* illustrates better than anything the complete opposition of the forest to the village. The pygmies consciously and energetically reject all village values in the forest. And when they are in the village they temporarily adopt its values and customs, not wanting to desecrate their sacred forest values by bringing them into the village. That is why they will never sing their sacred songs in the village the way they do in the forest, and why they refuse to consecrate the *nkumbi* with special music, although every other event of importance in their lives is marked in this way. There is an unalterable gulf between the two worlds of the two people.

CHAPTER TWELVE

The pygmies have their own way of growing naturally into adulthood. A boy proves himself as capable of supporting a family when he kills his first real game, and proves himself a man when he participates in the *elima*. During the whole *nkumbi* there was only one pygmy who showed the slightest fear of the consequences of the blatant disregard everyone else showed for the laws of the villagers. That was Aberi, father of Sansiwake. Aberi the ugly; Aberi who was so often the laughing-stock of his fellows. It was not that he was afraid of the villagers as people, for following his son's illness he often used to walk right among the instructors and drag his Sansiwake away, if he felt they were being too severe with him. And he was as outspoken as any against any mistreatment of the *baganza* by the villagers. He particularly criticized Sabani for his son's illness. Sabani and the other instructors told Aberi it was a terrible thing to do to criticize the *nkumbi* doctor, and not long afterwards Aberi became ill. He was so ill that he could not leave the initiation camp, and he slept on the wooden initiation bed beside the boys, moaning and tossing in his fever.

When he recovered he said that he had been bewitched, that the villagers were trying to kill him. He even attacked his brother Masalito for having contradicted just about every rule in the book. The pygmies practise no sorcery or witchcraft, but they know it is rife amongst their neighbours and they have seen enough of it in action to be uneasy about it. Aberi was one of the few who really believed in it, and who knew what real fear was. He knew fear not only of being bewitched, but of being accused of witchcraft; for his wife, Tamasa, the kindly woman whose face had been eaten away with yaws, was frequently being accused by the villagers. The villagers felt that anyone cursed with such a dreadful deformity must themselves be evil. On several occasions I heard pygmies accuse old Sau, the mother of Amabosu, the young singer, and say that she was a witch. But although they used the village word they meant something quite different by it. They meant no more than that she was being accused of making trouble. They sometimes said the same thing of other old people who were, as old pygmies often are, strong-minded and obstinate.

The villagers, however, believe that one can be possessed

of a substance which gives one supernatural powers, and compels the use of them to harm others. Sometimes it works in such a way that the witch may not even be aware of what he or she is doing, and various forms of divination are used to test suspects. The only time the pygmies think seriously about witchcraft is when they are in the village, or when villagers visit them in the forest. A single villager visiting a forest hunting camp can bring his evil with him, but when he departs it departs with him.

It so happened that Aberi, the one who feared this evil more than any, should be the one to die by it. He used to tell a story of how one night in the village he had been going to the latrine when he was caught from behind, as if by a hand clutching at his shoulder. But there was nobody there. He struggled and wrestled, but could not escape, and eventually he was thrown on to the ground. He suffered violent dysentery from that time onwards. But in spite of this illness, Aberi was still strong and healthy enough to be one of the most energetic, if not one of the most successful hunters. And so when he fell ill during a visit to the village, some time after the *elima* had ended, nobody thought much about it . . . except Aberi. He claimed that he had been bewitched, and he knew why. He said that many months ago he and two other pygmies had promised a villager from the BaNgwana village of Dar-es-Salaam that they would give him a lot of meat if he let them have a sack of rice. The villager, unwisely, gave them the rice but never got the meat. After several attempts to get his due he cursed all three with a powerful curse, and it was this curse, Aberi said, that was killing him. This villager not only possessed the witchcraft substance, but he knew it and used it deliberately. He was a sorcerer. Nobody believed Aberi, however, and slowly he got better.

One evening I met him wandering up the path leading from Cephu's tiny camp to our own, and I asked him how he was, saying that it was good to see him walking again. He often wore a dismal, dejected look; but this time his face was completely empty of any expression whatsoever. It was frightening to see a face normally so full of one emotion or another so void. Even his ugliness had deserted him, and in its place there was just nothing. Aberi said, slowly and carefully, that his body was better, but that he was going to die because he had been cursed, and for that there was no cure. He spent a short while in our camp, and we all tried to cheer him up. He seemed to brighten after a little singing, and when he left us to return he was almost happy.

Early in the morning, while it was still dark, I was awakened by an outburst of the most terrible wailing and crying, and I knew what it meant without asking an ash-grey Kenge who came pounding at my door. I went outside, and in the dim light I saw pygmies all over the place, standing in the doors of their huts, listening. One after another said, 'Aberi is dead, completely dead.'

Apparently he had gone back to his camp, and Tamasa had given him a little hot broth to warm him up, as he complained of being cold. He had gone to sleep, but later, quite suddenly and without any warning, he had awakened with a start, struggled violently as though trying to get up. But some unseen power was clutching at his shoulder, and Aberi gave a loud cry and fell back dead. Within a matter of weeks both the other pygmies who had been cursed with Aberi, had also died, under similar circumstances.

If the pygmies had believed in witchcraft, such a terrible series of coincidences would certainly have brought forth an accusation. As it was they never even mentioned any possibility other than that the villager might have used poison. This could have been in Aberi's case—but in all three? The pygmies shook their heads and said nothing.

Aberi was buried the following afternoon, with villagers officiating. Sansiwake was almost beside himself with grief, and Tamasa had to be dragged bodily away from the grave. When the women had all left the villagers started their usual inquiry into the death, to see who was responsible. They did not mention their fellow-villager from Dar-es-Salaam, the one who had cursed Aberi; instead they named Tamasa. Kelemoke and Masalito, who had their arms around Sansiwake and were trying to comfort him, turned around and looked at the accusers. Kelemoke spoke for them both. He said, quietly, but with considerable feeling, 'Tamasa is our sister, Tamasa is our mother. *We* do not kill our own kind.'

Without waiting for the villagers to finish their formalities the pygmies all left to return to the camps, and to make arrangements to leave the village and go back to the forest, where such things do not happen.

Chapter Thirteen

Towards the end of my stay at Epulu I felt that it was time to get away to other parts of the forest. I had been with the one hunting group for a year, moving about the forest with them, coming in with them on their trips to the village, returning with them to their hunting camps. It had been a full year, and the privilege of sharing the life of this group had begun to tell. Living so closely it was impossible not to begin to see the world as they saw it; sharing their pleasures and sorrows made it impossible not to feel as they did; and just being with a people so filled with the love of the forest made it impossible not to love as they did. The more I wanted to stay with them, and them alone, the more I knew it was time to move.

I asked Kenge if he would like to come with me, and was not surprised when he packed off Maliamo, sending her to stay with his mother. Nothing would have stopped Kenge from making our projected tour of the forest. We decided to circle the area by following the road around the fringes, stopping wherever it seemed expedient and setting off from there on foot. I took a supply of notebooks and pencils, a typewriter and a couple of changes of clothes, Kenge not only brought two changes of clothes, specially purchased at the local stores for the occasion, but also a huge old charcoal heated flat-iron. He said that we would probably be going through large cities and meeting big people, and he wanted to show them that we were not savages.

Our first stop was the headquarters of the BaNdaka Chief, at the village of Bafwakoa. It was a comfortable day's drive, and we arrived late in the afternoon, having stopped at almost every village on the way. The news of our coming had gone before us, in the strange way that news travels in Africa, from village to village. When we reached Bafwakoa, Chief Kachui was sitting out on his veranda waiting to greet us, surrounded by his official advisers and some of his family.

Kachui was an old man, and a leper, and he knew enough

of Europeans not to offer to shake hands. But we had met several times before, and he clasped both my hands warmly in his and made me sit down beside him. I introduced Kenge as being a friend, and with great aplomb the powerful old man called for another chair to be brought. This was the first time, I am sure, that such an honour had been done a pygmy by this severe and conventional chief. We talked a long while, and I made it quite plain that he could talk freely in front of Kenge. He did so. He said that he 'owned' several hundred pygmies, and they were really more trouble than they were worth. 'Here it is, time for the cotton crop to be brought in, and still I cannot get the pygmies to come back from the forest. They say the hunting is too good, but if it is, then why do I get no meat? What shall I say when the administrator comes and finds the cotton is not ready?'

I suggested that he could send one of his sons into the forest after the pygmies, guessing what the answer would be if pygmies here were like pygmies at Epulu. And sure enough Kachui shook his head and said he had sent three sons after his pygmies, but each time when they reached the hunting camp it was deserted, and nobody seemed to know where they had gone. Kenge shook his head and I was afraid he was going to laugh. But he merely said what a terrible way that was to behave, and that everyone knew that pygmies should always do what their master told them. This surprised Kachui, and he warmed to his theme, telling of all the ways in which he was constantly being tricked and cheated and robbed by the BaMbuti. Kenge continued to look disapproving, clucking his tongue and nodding in agreement even when Kachui said he would be justified in putting every last pygmy in jail—when he could find them.

But Kenge proved to be more than diplomatic. In turn he began questioning Kachui, and showed that over the past year he had come to know just the kind of topics I liked to hear discussed. And the fact that he raised the subjects dispelled all suspicions that Kachui might have had, and he even told us the exact whereabouts of such pygmies as he knew he could find for us, most of whom were tending his illegal plantations. He offered to call them in if we wanted to question them, but we said no, we would rather go and visit them. After some hesitation Kachui agreed, and made arrangements for us to leave the next morning. He installed us in his official guest-house, and sent over one of his sons with gifts of food. He also sent over two of his daughters, and I was tempted to find out whether they were both intended for me, or whether one was for Kenge. But I merely gave them a small gift each and said we were tired and were going to sleep. Kenge was thoroughly disgusted, and assured me that they did not have leprosy. I had a hard time trying to convince him that this was not the point, and he finally just rolled over in his blanket muttering that sometimes he thought Europeans were as crazy as villagers.

For the next few days we visited Kachui's plantations and talked with such pygmies as we found there. But it was impossible to shake off the Chief's sons, who stuck to us like leeches. As a result I managed to get no response at all, everything that the pygmies said was said to impress the villagers. The only useful information came from Kenge who got into the habit of slipping away while I was 'holding court', as he called it, and talking to his fellow pygmies well out of earshot.

But one thing was plain, and that was that no matter how powerful Kachui was—and it was certainly within his power to have any pygmies put in jail and flogged—he was no match for the people of the forest. Kenge learned that the main group had been away for over six months, and had no intention of returning yet. They were a long way off, and only came close to the village when they wanted to raid the plantations. Even while we were there Kachui's sons tried to find out where the rest of the pygmies were, but everyone protested complete ignorance.

A few days at Bafwakoa were more than enough. With Kachui lying in wait the pygmies were not likely to tell us where their companions were, and even if they did Kachui was not likely to let us out of sight. And he became unco-operative when I asked to look at the tribunal records, for which I had official permission. He was forced into admitting that he did not consider the pygmies as real people, and therefore they had no right to have cases tried in the tribunal. He said that any complaints by the villagers against pygmies were heard by him; he judged them himself and no records were kept. He made it quite plain that the judgment always went in favour of the villager. Kenge then disgraced himself by asking what happened if a pygmy brought a complaint against a villager. Kachui answered abruptly: 'Pygmies have no right to complain.' It was obviously time to go elsewhere.

On leaving Bafwakoa we headed down a little track that led to the banks of the Ituri River, also known as the Aruwimi. Across the river lay the territory of the BoMbo and BaBali tribes, both with unsavoury reputations for cannibalism amongst other things. The ferry was waiting on our side of the river, a series of huge canoes, each about thirty feet long, held together by a platform lashed across the centre, just wide enough and long enough to take a truck or two small cars. The men who operated the ferry were BaNdaka, and they came lazily down to the river from their village. They climbed into the ends of their canoes and cast off, each taking hold of a long, stout pole. As we drifted out the leader took his pole and struck it on the bottom of his canoe. It sounded like a great signal drum being beaten, and the echo rolled up and down the river with the force of thunder. The other men imitated their leader, then began poling slowly across, singing and chanting. The river at that point must be nearly a mile across, thickly forested down to the water's edge on each bank. It was vast and lonely. About an hour's walk from the far side is the spot where Stanley stood and watched the confluence of the two waterways, the Ituri and the Lenda, and remarked that this was one of the most beautiful sights in the whole of the Dark Continent. We went and stood on the same spot as the sun was setting, and I felt that this indeed was the heart of Africa.

Our stay among the BoMbo and BaBali was not much more fruitful than with Kachui, from the point of view of meeting or learning anything about the local pygmies. The Chiefs, Isiaka and Lukamba, both had the same complaints to make—that they simply could not control their pygmies. 'How can we control them?' Isiaka said. 'It is their forest, and they can hide from us whenever they want to. They are worthless people. They only come to the village when they want to steal.'

We did meet some pygmies, however, and they laughed and told us that no pygmy would ever come near Isiaka; there were plenty of other villagers who gave them much more in return for their meat, while Isiaka always expected them to work on his plantations for nothing. This, in fact, was the picture we got wherever we stopped. The villagers throughout the forest complained bitterly about the pygmies, saying, 'They eat us up until we are ready to die'; meaning that the pygmies take from them but give little in return.

It was not until we reached the western limits of the forest and turned north that we really had any luck. Some miles east of Wamba we left the car and followed a track through the forest. At the end of a full day we arrived at a series of old-established villages of the Mabudo tribe. They had a much more realistic attitude to the pygmies, and consequently did much better by them. They talked freely and willingly, and said that after all the pygmies were in the forest long before the Mabudo arrived, so they had the right to live as they wished. If they preferred to stay in the forest and live like animals that was their concern, but if they wanted to come to the villages and work for the Mabudo, that was fine. At the time of harvest the villagers all gave gifts of the first-fruits to the local pygmies, because, as they said, it was their land. And any time the pygmies were near a Mabudo village they invariably brought in gifts of meat or honey.

There were a few pygmies in one of the villages when we arrived, and after a couple of days they agreed to take us out to one of the hunting camps. Kenge had done some talking to them on the side, assuring them that I was not really like a villager, but just like them, a '*miki nde ndura*'—a son of the forest. When he introduced me to them he proudly pointed to the marks on my forehead that proclaimed me an adult. 'The villagers called him *Mulefu*, because he is tall,' he said. 'But we call him *Eba-mu-nyama*, because his father killed a buffalo, and he hunts with us in our forest.'

We left early in the morning. It would have been earlier if Kenge had not found an agreeable girl to sleep with during the night, and who insisted on cooking him an elaborate breakfast. But we finally got away about an hour after sunrise and headed eastwards. The ground was much more hilly than in the Epulu area, but the forest itself was much the same; tall and grand, lit dimly by the same filtered light. Underfoot the ground was even more clear of vegetation, as for the most part we were well above the streams and rivers. It was more rocky, and we passed a number of caves where our guide, an elderly pygmy with a serious face that kept breaking out into wrinkled old smiles, told us that we could pass the night if we were delayed, or if his friends had gone on to another camp. I did not much fancy the idea, and on looking inside one grotto I liked it even less; it was damp, and full of rotting leaves where some earlier travellers had made themselves a bed. There was barely room to sit up, and if it rained the water would certainly have drained into the cave.

We passed several trees of a kind I had not seen before, and Kenge said he had only seen them at the extreme northern boundaries of the Epulu district. The roots climbed into the air, gigantic, gnarled spirals, almost as thick as a man's body, joining the main trunk which towered above. Our guide told us that the Mabudo called this the 'elephant tree' because they always took refuge there if they were attacked by elephant. He dodged inside one to show us, hiding behind the roots. Even a Mabudo would have been able to stand upright with perfect ease. But the roots were close together, and an elephant could only get its trunk between them, he explained. And then the Mabudo sliced it off with their machettis. He imitated the action, laughing at the ridiculous picture he presented, trying to be the elephant and the Mabudo at one and the same time.

Kenge was amazed, and asked if the Mabudo actually came that far into the forest. The old man said that they did, but not very often. They came to dig traps or set up spear-falls for elephant. They both laughed at the strange hunting methods of the villagers. The digging of a trap was a long and arduous task, as was the setting up of a spear-fall. The latter method is particularly difficult. A stout spear blade is hafted into a heavy tree trunk which then has to be hauled up high above the ground. A vine is stretched underneath and acts as the trigger to release it. But the animals soon come to recognize any kind of trap and simply avoid them. We passed several, and sure enough we could see the tracks going right up to the edge, then carefully circling around in safety.

Kenge could not get over the notion of hiding among the roots of a tree instead of going out and spearing the elephant as any pygmy would. For the rest of the day whenever we passed an elephant tree he would dodge in among the roots while the old guide pranced around outside pretending that he was the elephant, stooping down and swinging one arm loosely in front of him as though it were an elephant's trunk.

This slowed down our progress, and in the late afternoon our guide suddenly put on pace. It was nearly dark when we climbed up a high ridge and followed it for a mile or two. At the end, before descending, we stopped and listened, and from down below, to the left, we heard the unmistakable sounds of a pygmy camp: chattering, laughing; children playing and babies crying; and here and there someone singing. It was like being home again, and Kenge was all for going straight down. But our guide made us wait while he went ahead and warned of our approach. It would have been perfectly possible for the camp to have been deserted by the time we reached it if they had thought that we were in any way undesirable company. As it was, when we finally appeared, they all stood around in a silent and curious circle to welcome us, each family outside its hut, uncertain and shy.

For the next few days we stayed there, and then we moved on from camp to camp, visiting the six or seven hunting groups in the neighbourhood. Towards the end we found that we were only a day's march away from Kaweki's fishing camp, and another day would have brought us back to Epulu. So we turned about and slowly retraced our steps. Once we had become known to the first group we were accepted wherever we went as a matter of course.

Kenge, who had been extremely shy and rather nervous in the Mabudo villages, refusing to eat the food they offered him, was once more his old self. He exploited to the utmost what was, for him as well as for our hosts, a novel situation. Generally pygmies only travel within their own area and amongst groups that they know and where they have relatives. There was a great deal of curiosity on both sides, but it was mainly concerned with the villagers, what kind of people they were and how easily they were tricked. The pygmies that we were with for the most time in this region only visited the villages rarely, and some of them had never seen a European before, and none of them had met a pygmy from the other side of the forest. They examined both of us with attention. They pitied me for my height, which made me so clumsy, and they laughed at the hair on my arms and legs, which they said was like that of a monkey. A pink face poking out from behind a beard added to the similarity, and it became a standard joke amongst them, much to Kenge's annoyance. He felt that it lowered his own status.

For me the trip was mainly work; visiting different groups and taking copious notes among each. But for Kenge it was an experience beyond all imagination. Not only was he able to meet other groups of his fellow BaMbuti, and many different tribes of villagers, but he saw different kinds of Europeans . . . missionaries, business men, movie-makers, 'white hunters', and others. We visited the town of Paulis; it was only a small town, but the biggest that Kenge had seen, and with a railway which was something that aroused his deepest curiosity. He could not think why the trains were limited in their movement by the iron rails, and could not turn wherever they wanted to. There was something about this restriction of freedom that bothered him. I took him to see a football match between the workers of a local brewery and those of a large combine store. Just prior to the match we had been drinking palm wine in a small African bar, and found it to be much stronger than the wine made in the forest. So Kenge rooted happily for what he called 'Men of the Beer', but just why people should spend the afternoon kicking a ball backwards and forwards was as puzzling to him as were the captive trains.

We only spent a few days in Paulis, branching out from there along the northern limits of the forest. Then we turned eastwards and made our way across and down into Lese country. The BaLese are considered to be arch sorcerers and witches, and it was with great reluctance that Kenge agreed to stay in the village of Chief Lupao, near Nduye. It was several miles off the road, and by accident or design it was empty when we got there. Fires were still smouldering and smoking, but there was no other sign of life. The Chief's *baraza*, a huge edifice with a thatched roof, stood open in front of his house; the fire had been recently made up, but nobody came out to greet us.

Kenge was getting more and more nervous, and I was on the point of giving in to his repeated requests to leave and find a more hospitable place to stay when a young man appeared and wandered idly towards us. He stopped some way off, and stood there looking at us. This was a completely new situation to me as in every other village, even where we had been total strangers, we had been almost embarrassed by the hospitality of the people. I waited to see if the man's curiosity would get the better of him, but after a while I gave up and went over to talk to him. Kenge refused to move from the safety of the car, and stood there with his hand on the door, ready to lock himself inside at a moment's notice. For him this was a foreign world indeed, for these were not only villagers, but villagers who had a name for devouring corpses.

I made little headway with the young man, who continued to chew on a stick while he avoided giving any direct answers to my queries as to where Chief Lupao was. He questioned me in turn, and when I had assured him that I was not a missionary, nor a government agent, not a plantation owner looking for forced labour, he wandered away . . . still chewing. He called back over his shoulder that he would try and find his father. Kenge was speechless with indignation that we had not been invited into the *baraza* or offered water to wash with. He said that even villagers did that much for strangers, however unwelcome. He was more unhappy than ever.

Eventually Lupao's son returned and said that his father would be back soon from his plantation, and we could sit down in the *baraza* if we wanted to. With that he walked away and once more we were alone. By now it was too late to think of going anywhere else, so we entered the *baraza* and sat down on a couple of logs near the fire. After an hour a small crowd of old men and children had gathered around and were staring at us, silently and speculatively. Any time either of us tried to talk with them they just shook their heads. Kenge was beginning to be more impatient than afraid, and he demanded that someone go and fetch water for us to wash with, asking the bystanders what kind of savages they were to let their guests stay unwashed and unfed. The Chief's son looked at Kenge coldly and asked him what tribe he belonged to. Kenge replied without hesitation that he was not of the village, but of the forest . . . he was a pygmy. The young man nodded and said, 'So I see!' And still no move was made to fetch us any water.

Just then Chief Lupao came out of his house, an extremely surly and untidy old man who did not try to disguise the fact that he had been sleeping, and not in his plantation. He greeted us in an off-hand way and asked what we wanted. I explained that we wanted to go into the forest to visit some of his pygmies, and had come to pay our respects first. He said brusquely that he did not know where any pygmies were to be found. Then he looked at Kenge and asked him if it was usual for a pygmy to sit beside a white man. Kenge countered by asking Lupao if it was usual for a big Chief to keep a white man waiting while he slept, and to leave him sitting on a log without so much as a glass of water to drink or to wash with. At that the other villagers began joining in, and feeling more than uncomfortable at being made the centre of a fight between them and Kenge I told Lupao that we were very tired, and asked if it would be all right for us to stay the night in his village. Lupao, who by then was engrossed with a steaming bowl of chicken soup, said that he supposed so, and indicated with a nod of his head for his son to go and find us a place to sleep.

I made a few more attempts to make polite conversation, but did not get so much as a word in response. Everyone was silent, staring either at us or up at the smoky roof above us. It had been quite dark for some time, and after waiting to see if the son was going to do anything about sleeping quarters I got up and said loudly to Kenge that I was going to sleep in the car, and asked him if he was going to do the same. He replied that he would sleep outside to guard me from all the thieves and murderers that made up the entire population of the village. This was putting it a little more strongly than I had intended, and we had now given real cause for offence. As we left there was a stony silence, and Lupao refused to look up and allow me to take my leave of him. I can hardly blame him, but the tradition of hospitality among most African villagers is so strong that one comes to accept it and to expect it. There are invariably some responsibilities one has to meet in exchange, but with the BaLese it seemed otherwise.

When Lupao saw that we were really leaving in such an un-ceremonious huff, he called his son, who had remained in the *baraza* all the time, and sent him running after us to tell us that a house would be ready soon. Kenge replied that the car was clean, which was a little too subtle for the young man. He re-turned half an hour later with some food, and this time Kenge was more direct. He lowered the window of the car enough to say, 'We people of the forest don't eat excrement!' and then hurriedly wound it up. We passed an uneasy night, and in the first glimmer of morning, without as much as looking for a stream to wash in, we left. Nobody even came to watch us leave.

We were so tired and exhausted that we decided to call on the Catholic Mission of Nduye and at least get cleaned up. We drove through the little village, and on the outskirts turned on to a steep stone track that wound up the hillside. It climbed erratically past a temporary church, near which rough-hewn stones were being assembled for the building of a more per-manent structure, and on to a ledge just wide enough to hold the long, low mission house and a neat terrace in front, where the ground dropped away sharply. The welcome we got there was very different. A huge, bearded figure in white robes came striding up the hill after us, shouting instructions in KiNgwana to cook up another half-dozen eggs. To us he said, 'You are late for breakfast.' And after a second look, 'You need hot baths.'

He would not hear of our going anywhere else, and insisted that we stay for at least a week, a month if we liked. There was plenty of work we could do and we looked as though we needed to go to church. Then he asked me if we were Catholics, and when I replied that I was a not very good Protestant and that Kenge was a practising heathen he roared with laughter and said that at least Kenge didn't know what a mistake he had made. His good humour put Kenge immediately at ease, and when Father Longo called for his Lese Catechist to show Kenge where he could wash, eat and sleep, there was only a momentary flicker of doubt. Shortly afterwards I heard Kenge in the distance, telling his new companions in his loudest and most boastful voice of what a terrible experience we had been through the night before. As for myself, I found myself in the company of the kindest, wisest, and most sincere person I had met in this part of Africa.

After he had seen that I did justice to an enormous breakfast he took me down to look at the new church he was building. Muslims and Pagans worked side by side with Christians, breaking rocks on the mountain-side, carrying them to the site, squaring them and heaving them into place. All this help was voluntary, Father Longo paying the workers whenever he was able, which was all that was expected. It was done willingly because over the years the people of the surrounding villages had come to know Father Longo for the man he was. He was not concerned simply with adding a lot of names to his role of converts, but primarily with living a life that would be an example that others would want to follow. He made everyone equally welcome at his mission, regardless of their religion. Some months later, when the new church was finished, it was consecrated at a service made all the more beautiful by dozens of his pygmy friends who came in from the forest carrying gifts of forest fruits and flowers, dancing their pagan way up the aisle. They were filled with as much joy and devotion as Father Longo himself.

We came to know the pygmies in the neighbourhood of Nduye well, but in spite of our initial set-back with Chief Lupao we still wanted to make some contact with the BaLese. Father Longo suggested that we try the other powerful Lese Chief, Nakubai, whose village was a number of miles to the north. We followed his advice, and Nakubai did everything he could to reverse our opinion of his people. He was not only hospitable, but he understood when I said that we wanted to visit his pygmies without the restraint of having a villager along with us. Whereas Kachui and others had insisted on keeping us under constant surveillance Nakubai even said that we would probably see far more if we went on our own. As a result we had one of our most fruitful treks of all, learning many new and important things about the forest people.

The terrain was very different, being mountainous and

rocky, jagged boulders sticking out of the ground at crazy angles, as though some great city had once been swallowed by an earthquake. But even on the steepest escarpments, where the huge trees seemed to grow out of the very rocks, the pygmies moved as quickly and surely and silently as elsewhere. For the first time I saw Kenge in difficulty, he being as unused to this kind of mountaineering as I was. But we were anxious to press inwards as far as we could, and once again we found ourselves almost back at the Epulu; it lay about two days to the south-west at the most.

While we were in this part of the forest we heard many stories of the old wars between the various tribes of villagers, as they invaded the forest, and how the pygmies got caught between them so that they were forced to side with one group or another.

On leaving Nduye we continued southwards, past the administrative centre of Mambasa, and when we had crossed the Ituri, which at this higher point was, in Kenge's words, 'a mere child', we once more left the car and began walking. At the end of a day, having passed through three deserted pygmy villages, we were beginning to feel dubious at the wisdom of having come so far without a guide when we came to the banks of a large tributary of the Ituri. And there, to our mutual surprise, was one of the legendary vine bridges, slung high up over the swirling waters, suspended by lianas fastened to the topmost branches of the trees on either bank. I was glad that Kenge was amazed, because there is every reason to believe that these bridges are, contrary to the information supplied by romantically minded movie-makers, *not* of pygmy origin. To cross the bridge we climbed up a series of logs that led to a fork in the tree to which our end of the bridge was anchored, and from there we made our swaying way across. Kenge refused to look down, and when I stopped in the middle to get a better view of how the supporting lianas were fastened, he urged me on, saying that it was not natural to be standing there in the middle of the air over a dangerous river.

On the far side we came to a camp, and there we found some relatives of the Epulu pygmies, including one girl that was well-known to both Kenge and myself—Makubasi's first wife.

We stayed there for one night, and then moved on up a trail that led to the Protestant mission of Biasiku. This was headed by an American missionary, the Rev. Charles Bell, and his wife. They had been in the Congo for about as long as Father Longo and Putnam. As soon as they heard that we were staying in the camp down by the river, they made arrangements for us to stay in the mission, and without knowing that we were on our way up Mr. Bell set out on foot to issue the invitation in person. I had a special reason for accepting it; I wanted Kenge to have a fair notion of the way different missions operated. I had no desire to see him converted, but having given him a taste of missions it seemed fair that he should have an all-round picture, and earlier during our trip we had had an unfortunate experience with a different Protestant mission.

At that mission prayers were said in public about five times a day, mention always being made of any converts who might be ill, however slightly. Prayers were also made requesting the Almighty to provide a road into the forest so that the mission could have easier access to the pygmies, and thanks were offered for a new automobile that had been acquired. All this was perfectly understandable to the villagers, with their own tradition of witchcraft. They merely thought this a particularly powerful form of magic. Kenge was sceptical. 'How is their God going to put a road into the forest when the white man can't do it?' he asked.

But one day, while we were there, a wounded pygmy was brought in. He had been gored while hunting, and was groaning with pain. The missionary and I were sitting outside at the time, going through his morning ritual of drinking tea. I went over to see what had happened, and if they wanted any medicine. They said no, but they would like something to make the pain go away until they could prepare their own medicine. All I had was aspirin, but in any case I had more faith in their powers of healing than mine. Kenge asked in a whisper why the missionary had not come over, but was continuing to sip his tea, keeping his back to us. I did not know, and I said nothing until the next prayer meeting when, as usual, prayers were offered for all those converts who had minor ailments, and for the new road. But not a word was said about the pygmy, lying in a hut not more than ten yards away. Then I did ask, and I was told, 'Why should I pray for him? He is not a Christian.' The pygmies were all standing around, and heard, as they were meant to hear. They had little idea of what Christianity was, but from that moment I doubt if any of them think that it has much to offer them. When, at the end of our stay there, the missionary asked Kenge what he thought of the Christian God now that he had been privileged to live in a Christian community, Kenge said with his usual frankness, 'It is the biggest falsehood I know.' And so I was glad for Kenge to meet other Christians who had a better idea of what their religion stood for, and the Bells were certainly such.

It was while we were there that Kenge first began to realize that there really was another world, beyond the forest. He had often heard people talk about it, and up at Paulis the forest was thinned almost to mere savannah, but nowhere had he really seen beyond the forest. The mission was built up on a high hill, cleared of vegetation so that crops could be planted to supply the mission school and hospital with food. Usually the distant horizons were hazy, and everywhere we looked down only on a sea of trees. Several times I tried to tell Kenge that beyond this lay another world, a world of lakes and mountains and open plains, but he just nodded disbelievingly. He had never even seen a hill, as such. He had climbed up them, inside the forest, and had seen small hills such as the one on the far side of the Epulu, and the one on which the Nduye mission stood, but that was all. One evening, however, we were both sitting down on a boulder at the highest point of the Biasiku mission, looking over the forest. To the east the sky was becoming less and less hazy, and gradually out of the haze a great black mass began to form. Kenge saw it before I did, and asked if it was a bad storm cloud. It was the vast bulk of the Ruwenzori, the Mountains of the Moon.

We never saw the peaks, but even the lower ranges towered so far above the forest that Kenge was almost frightened. He asked me about them, and I did my best; but it was next to impossible to describe things of which he had no experience, and for which there were not even any appropriate words in his language. His curiosity was thoroughly aroused, and feeling a slight hankering myself to see this lovely part of Africa again I asked him if he would like to have a holiday and go over there, to see the outside world for himself. He hesitated a long time, and asked how far the mountains were beyond the forest. I said that they were not more than a day's drive from the last of the trees, to which he responded with disbelief, '*No* trees? *No* trees at all?' He was not at all sure about this, and asked me if it was a good country, and what the villagers were like there. I told him that it was very much like my own country in some ways, that it was very good, and that there were few villages.

This decided him and he agreed to go, providing that we brought enough food to last us all the time we were to be out of the forest. He was going to have nothing to do with savages who lived in a land without trees.

Chapter Fourteen

WE STOCKED up with food; stalks of plantains, a basket of rice, a large bundle of manioc roots, some dried fish and a large quantity of peanuts, with which I knew Kenge would make a delicious sauce together with some forest tomatoes, onions and peppers. In case I could not persuade him to use local water for coffee, all I could think of in the way of acceptable liquid was beer.

We drove on down to Beni, where the forest ends, passing through in a violent storm that cut the visibility down to a few yards. It was almost impossible to see the edges of the road, and Kenge was not even aware that we had left the forest. But he noticed that we were climbing, and that it was getting colder and colder. He shivered and pulled my blanket around him, keeping his eyes ahead, not looking to one side or the other. After a while there was a patch in the swirling mist, and to our left I could see the side of one of the foothills, rising sheer and black, soon lost again in the clouds. Kenge shook his head and asked if that was the hill we had seen from the mission station. He immediately began talking about the mission, and how pleasant it had been there, and how nice it would be to go back . . .

The rain closed in again, and it was only when we were almost at the entrance to the Ishango National Park that it cleared. We stopped at the check point and paid the small fee that entitled us to stay overnight in one of the small guest-houses, and to the services of the guides, who compulsorily attended every traveller through the park, for the protection of both travellers and game.

Between us and the guest-houses were some twenty-four miles of park, traversed by narrow roads made thick with mud by the recent storm. Kenge moved over and sat beside me, making room on the outside for the burly guide, an Azande named 'Henri'. Before us the road rose up a slight incline, and all around was nothing but rolling, grassy plains. The mountains were directly behind us. Kenge kept on muttering, 'Not a tree, not a single tree—this is very bad country.' But I reassured him that it was very good country, and that he was going to see lots of game. At this he perked up, and Henri saved the situation by asking Kenge if he was a pygmy, and then telling him that he would see more game than he had ever seen in the forest, but he was not to try and hunt any. Kenge could not understand this, because to his mind game is meant to be hunted, and he started a long argument with the good-natured guide who soon had Kenge laughing happily.

As we got to the top of the first incline, there was a flat stretch and then a steep rise. Half-way up this the car sank down into the mud and would not move. Henri and Kenge got out and started pushing, and after half an hour of hard work we reached the top. Once the wheels began gripping I did not stop but kept going until I was at the very crest. There I stopped the engine and waited for the other two who came running up the hill after me, joking and shouting, covered from head to foot with smelly black slime.

When Kenge topped the rise, he stopped dead. Every smallest sign of mirth suddenly left his face. He opened his mouth, but could say nothing. He moved his head and eyes slowly and unbelievingly. Down below us, on the far side of the hill stretched mile after mile of rolling grasslands, a lush, fresh green, with an occasional shrub or tree standing out like a sentinel into a sky that had suddenly become brilliantly clear. And beyond the grasslands was Lake Edward—a huge expanse of water disappearing into the distance—a river without banks, without end. It was like nothing Kenge had ever seen before. The largest stretch of water he had ever seen was when we had stood, like Stanley, and marvelled at the confluence of the Lenda and the Ituri. On the plains, animals were grazing everywhere; a small herd of elephant to the left, about twenty antelopes stared curiously at us from straight ahead, and way down to the right was a gigantic herd of about a hundred and fifty buffalo. But Kenge did not seem to see them.

As he turned around to look behind, I turned with him, and even I was overcome with the beauty of what we saw. It was one of those rarest of moments that come perhaps once or twice a year, and may last for only a few seconds, when a violent storm clears the air completely and the whole mighty range of the Ruwenzori mountains is exposed in a wild and glorious harmony of rock and snow. The lower slopes shone green with the dense forest; the upper slopes rose in steep, jagged cliffs, and above them the great snow-capped peaks rose proudly into the clearest possible sky. There was not a cloud to match the pure whiteness of the snow. Kenge could not believe they were the same mountains that we had seen from the forest—there they had seemed just like large hills to him. I tried to explain to him what the snow was—he thought it was some kind of white rock. Henri said that it was water that turned that colour when it was high up, but Kenge wanted to know why it didn't run down the mountain-side like any other water. When Henri told him it also turned solid at that height Kenge just gave him a long, steady look and said, '*Bongo yako!*'—'You liar!'

With typical pygmy philosophy, even though he could not understand it he accepted it, and turned his back on the mountains to look more closely at what lay all around him. He picked up a handful of grass, tasted it and smelled it. He said that it was bad grass and that the mud was bad mud. He sniffed at the air and said it was bad air. In fact, as he had stated at the outset, it was altogether a very bad country. The guide pointed out the elephants, hoping to make him feel more at home. But Kenge was not impressed—he asked what good they were if we were not allowed to go and hunt them. Henri pointed out the antelopes, which had moved closer and were staring at us as curiously as ever. Kenge clapped his hands together and said that they would give food for months and months. And then he saw the buffalo, still grazing lazily several miles away, far down below. He turned to me and said, 'What insects are those?'

At first I hardly understood, then I realized that in the forest vision is so limited that there is no great need to make an automatic allowance for distance when judging size. Out here in the plains, Kenge was looking for the first time over apparently unending miles of unfamiliar grasslands, with not a tree worth the name to give him any basis for comparison. The same thing happened later on when I pointed out a boat, in the middle of the lake. It was a large fishing boat with a number of people in it. Kenge at first refused to believe it. He thought it was a floating piece of wood.

When I told Kenge that the insects were buffalo, he roared with laughter and told me not to tell such stupid lies. When Henri, who was thoroughly puzzled, told him the same thing, and explained that visitors to the park had to have a guide with them at all times because there were so many dangerous animals, Kenge still didn't believe, but he strained his eyes to see more clearly and asked what kind of buffalo they were that they were so small. I told him they were sometimes nearly twice the size of a forest buffalo, and he shrugged his shoulders and said he would not be standing out there in the open if they were. I tried telling him they were possibly as far away as from Epulu to the village of Kopu, beyond Eboyo. He began scraping the mud off his arms and legs, no longer interested in such fantasies.

The road led on down to within about half a mile of where the herd was grazing, and as we got closer, the insects must have seemed to get bigger and bigger. Kenge, who was now sitting on the outside, kept his face glued to the window, which nothing would make him lower. I even had to raise mine to keep him happy. I was never able to discover just what he thought was happening, whether he thought the insects were changing into buffalo, or whether they were miniature buffalo growing rapidly as we approached; his only comment was that they were not real buffalo, and he was not going to get out of the car again until we left the park.

We continued down towards Lake Edward, blue and clear and smooth. The shore line stretched across the wide plain to the east, where the distant hills of Uganda rose and shimmered uncertainly until they were lost in the south. The lake itself disappeared into the southern haze, and all along the west it hid itself in the low-lying plains and swamps beneath the hills of the Congo. The Semliki River, which joins Lake Edward with Lake Albert, to the north of the Ruwenzori, cuts through the plains in a wide gulley, flanked by cliffs some two hundred feet high. At the top of the cliffs, just above where it begins to drain from the lake, are the guest-houses; two neat brick sleeping quarters, each divided into two sets of rooms, with a dining-room and kitchen building in between. About a mile to the east is a small village which houses the African staff and guides, and where the servants of visitors can stay.

The guest-houses face the edge of the cliff, which is not much more than a hundred feet away, and each has a built-in garage and instructions to leave the car inside at all times when not in use, in case of visitations by elephant, buffalo or hippopotamus.

Both the guide and guest-house staff were surprised when I said that Kenge would stay with me in the guest-house, and not go to the village, but Kenge had made it quite plain that he was not going to let me out of his sight for a minute. And besides, I wanted to see his reactions. He was considerably cheered at the sight of comfortable living quarters, and after having what was probably the first shower of his life and changing into clean clothes, he promptly set off for the kitchen to cook a meal. The official cook was put out by this, so I said that I would pay him anyway, but that I would rather let Kenge cook our food. This was all right, as it meant the cook had nothing to do but sit and watch. But when Kenge brought two plates into the dining-room, the cook protested violently, saying that *indigènes* were not allowed to eat in the dining-room. I said that Kenge was my friend, and the cook replied that might well be, but that if the Belgian park superintendent came around and saw an *indigène* eating in the dining-room he would lose his job. I asked if it was all right for Kenge to eat in the kitchen, and he said, yes, that was quite in order, so long as he did not use the guest-house crockery. So we got our own cooking utensils and enamel bowls, and sat down on the floor of the kitchen. The cook protested even more loudly at that, and called the village headman over. Kenge started talking to me in KiBira so that they could not understand, and the headman realized he was getting nowhere. He simply posted guards to warn him if the superintendent showed up.

Before it grew dark we had time for a quick drive around part of the park, but except for coming across a family of elephants crossing the road right in front of us, we saw no game. The guide said the elephants were on their way down to the Semliki, and that if we went straight back to watch we would see them and all the other animals as they came down to drink and bathe. I had visited Ishango a few years earlier, and remembered sitting on the edge of the cliff and watching the animals on the far side of the river come slowly across the plain and down the narrow, steep, cliff path to the water. Henri led us both to a different spot, sheltered and hidden by clumps of gigantic cactus plants, and told us we would see more from there.

But first Kenge wanted to wander along the top of the cliff, to see if it was really true that during each night the hippopotami clambered up to the top and 'danced' in front of the houses, as Henri had said when he warned us not to set foot outside after dark. I had slept outside once before and would never do that again; and sure enough as Kenge walked along, carefully scanning the ground, he found tracks of almost every kind of animal, including hippopotamus. We returned to our clump of cactus and sat down. For the first time, Kenge spoke approvingly. Perhaps he was reassured because to the far south the great Kivu range of volcanoes had appeared out of the haze, putting an end to the lake. Behind us the Ruwenzori were swathed in a ringlet of clouds half-way up, but the peaks above shone white in the darkening sky. The air was still, and the only sounds were those of the lake birds crying as they swooped low over the water. I felt, all of a sudden, as though I was back in the forest, and when Kenge spoke I knew why. I only wish I could somehow convey the sound of his voice— soft and musical, so quiet that I could hardly hear it. I have seen many of the forest people talk like this when they are feeling deeply. Even the movement of his mouth and his eyes and gestures of his hands conveyed what he was saying. As if singing to himself, he said, 'I was wrong, this is a good place, though I don't like it; it must be good, because there are so many animals. There is no noise of fighting. It is good because the sky is clear and the ground is clean. It is good because I feel good; I feel as though I and the whole world were sleeping and dreaming. Why do people always make so much noise?' And then he added, with infinite wistfulness, 'If only there were more trees . . .'

As the sun sank lower we watched the hippopotami playing

in the Semliki, directly beneath us. They heaved their great bodies out of the water and opened their gaping mouths wide, snorting and grunting before they sank back beneath the surface with a tremendous splash that sent waves rolling to the shore. Further down the river two bulls were fighting, but only half in earnest. Soon they all subsided and only the tips of their noses were left poking just above the surface of the water.

Then came a procession of almost every type of animal in the park, one after the other, each picking its way with unhurried grace to the water's edge. On our side of the river they came right past where Kenge and I were sitting, down a narrow path that disappeared from sight beneath us. We next saw them as they emerged at the bottom and carefully waded into the water, looking all around to make sure that there were no rivals nearby. Then, as one group of animals moved away another came and took their place. The elephants that we had passed on the road were the last that we saw, and as it was getting dark, we retreated to the safety of the dining-room, where we cooked and ate our evening meal in the kitchen, leaving the dining-room unused, for there were no other guests in the park at this time.

When we had finished, Henri tried once more to persuade Kenge to go with him to the village, but Kenge refused. Warning us to go straight back to the guest-house Henri disappeared, spear in hand, running lightly. We made our way to our house and sat out on the veranda for an hour or two, watching and listening. The elephants returned from the valley and walked right past us, between the house and the cliff, so close that we could smell them and hear their snuffly breathing. Not long afterwards a single buffalo snorted directly to our left, and we could see its huge bulk at the edge of the garage, sniffing the mud-spattered motor-car. Other sounds came and went, but the night was dark and all we could see were vague shapes moving uncomfortably close at hand.

Then the cloud must have lifted, for the two beacons I had been waiting for gradually came into sight, dull patches of red glowing unevenly far beyond the cliffs in front of us. They were the two active volcanoes just to the south of Lake Edward, high up between Edward and Lake Kivu. I pointed them out to Kenge, who had heard tales of them, and was surprised that they should glow so plainly at such distance, while during the day he had only caught a glimpse of the far coastline. But just then the ground all around us seemed to shake, and there was the unmistakable grunting of several hippopotami. They must have been all around the house, and as they heaved their heavy bodies about in search of fresh grass, they made the earth tremble. It was no wonder the local guides referred to this cliff-top as 'the dancing ground of the hippopotamus', because as they grunted and thudded clumsily around it seemed that there was a curious rhythm to their movements.

When we looked at the ground the next morning, still soft from the recent rain, it was churned up by hundreds of different footprints, including the spoor of a lion, rare in this area, which led directly past the veranda steps where we had been sitting.

The morning was spent going all over the park with the now thoroughly friendly Henri who took us to places where he knew we would find different kinds of animals. I left him to deal with Kenge, and it soon became a game between them to see who could spot a tell-tale movement first. Kenge's eyes were sharper, but he was seldom able to identify anything that was more than a few hundred yards away, whereas Henri could tell at several miles.

Everywhere the view was different, and Kenge was as fascinated by this constant change as he was by the abundance of game. The sun shone warmly in a clear sky, but the Ruwenzori had hidden themselves once more in a bank of cloud. Looking at them from the north side of the park we could see only the forest-covered lower slopes. All at once Kenge said, 'I dreamed of my mother last night, and of my wife, Maliamo. My mother was crying for me.' I asked him if he wanted to stay another day or so in the park, or go back, and he said he wanted to go away that day, as soon as he could. When we got back to the guest-houses we heard that the superintendent was on his way, so we decided to leave at once rather than have any arguments as to whether it was better for a white man to eat in the kitchen with a pygmy, or a pygmy to eat in the dining-room with a white man. Henri returned with us to the entrance of the park, where he and Kenge said a fond farewell, Kenge giving him some of our stock of rice as a present.

Where the road leaves the heights above the plains and crosses over to the forest plateau, we stopped the car and got out to have a last look. It was at its most beautiful, and I waited for several minutes while Kenge took it all in, shaking his head and clucking his tongue in amazement. When he spoke all he said was, 'The *Père* Longo was right, this God must be the same as our God in the forest. It must be one God.'

*　　　*　　　*

Our homeward trip was a triumphant one. Kenge sang loudly as we got further and further into the forest. He was so happy to be back that he forgot temporarily about his mother and Maliamo, and himself suggested stopping in several places, which we did. Our last stop was outside Mambasa, where the local administrator in a fit of well-meaning but not very far-seeing zeal was 'liberating' the pygmies from the villagers. He did this by cutting down clearings along the roadside and setting up villages for the pygmies to live in, so that they could cultivate plantations that would provide additional crops for the government to collect and market. Kenge was immensely amused at the thought of being liberated, and said the pygmies would still go to the negro villages, why should they work on their own plantations when they can steal from real villagers?

We accompanied the administrator, a fat and jovial man, hospitable and friendly, who genuinely liked the pygmies. But he could not be made to believe that what he was doing was wrong. He had called various groups of pygmies down to the clearings, and he talked to each of them explaining that the government would supply them with all the necessary tools for cultivations, and would give them the seeds to plant. In time, each pygmy village would have a *baraza*, like the negro villages, and a school house; and the important ones would have their own dispensaries. He then asked in a democratic way if they wanted to come and join the plan. They all shouted yes, to a man, and the administrator turned to me with a victorious smile of pride. He then commanded the villagers he had brought with him to distribute beans to each pygmy, with instructions as to how and where to plant them. The pygmies all lined up, and as soon as they had received their share they rejoined the end of the line. When they were caught out and taken off the line to their allotment to plant the beans, they listened gravely and began to work. Half an hour later most of the beans were being cooked for lunch.

The plan was doomed to failure for several reasons. For one, the pygmies are not able to stand the direct sunlight, and become ill. They also become ill because they do not have any resistance, as do the villagers, to the kinds of disease they are open to in a sedentary life. Water which the villager can drink with impunity gives severe stomach disorder to the pygmy used to the fresh, clean water of forest streams. Flies and mosquitoes carry germs unknown to him in his natural home. But above all, his entire code of behaviour and thought is geared to his nomadic forest life; to bring him to a settled life in a village was to ask him to abandon overnight one way of life, a way he had lived for thousands of years, and adopt another. Where the pygmies did make the attempt there was complete moral as well as physical disintegration. Not long after a Belgian friend over from the Congo told me that the day before he had left, on one small 'model plantation' where the pygmies were being liberated, twenty-nine had died in one day—from sunstroke.

I spoke about the plan to many pygmies, and when Kenge got back to Epulu he talked as much about it as about anything else that he had seen on his journey to the outside world. Old Moke summed up the attitude that I had found everywhere. 'The forest is our home; when we leave the forest, or when the forest dies, we shall die. We are the people of the forest.'

Chapter Fifteen

OUR RETURN to Epulu found most of the pygmies, except for Cephu and his group, down at the village. Hunting had been exceptionally good and they had dried a quantity of meat over the past weeks and were now busy trading it for village produce. Their roadside village camp had long been abandoned, and only my house there was still standing. Everyone seemed to have agreed that they did not want any more mud houses, and they had made a temporary camp just inside the forest. It was not far from one of the plantations, and close to a stream of crystal water.

Kenge had a hero's welcome; even the villagers came to visit the camp to hear his stories. But none of his stories fascinated the pygmies as much as those of the model pygmy villages we had passed on our way home. He was made to repeat these accounts over and over again, giving every detail. While the younger men and women laughed until they could laugh no more at Kenge's description of the distribution of beans for planting, and how most of them ended up in the cooking pot, the older people were less amused. 'But are they actually living in those villages?' they asked. And when Kenge answered that some of them were, the next question was, 'What if there is no game nearby?' Kenge could only tell them what the administrator had said . . . 'You are welcome to hunt so long as you continue to look after the plantations.' Everyone knew that to be impossible. Hunting is a full-time occupation, and it could lead a group away for months on end. It called for the co-operation of men, women and children. Who was to stay to look after the fields?

Other questions brought equally unsatisfactory answers, even as far as the younger people were concerned. There were no trees in the 'model' villages . . . every last one had been cut down, exposing the huts to the full heat of the sun; the pygmies were to be supplied with free seeds to plant, but they would be expected to produce a definite yield, and they knew that villagers who failed to come up to the mark were fined, and if they were unable to pay their fines they went to jail. Every village had been placed near a supply of fresh water, tested by the administrator himself (by looking at it—he was horrified when I drank some) but the pygmies knew that however clean it might be to start with, it would soon be polluted, and in any case there was no guarantee that it would continue throughout the year. And far from being liberated from the villagers the pygmies felt that they would be in greater danger than ever. While in the forest they knew that they were safe, because the forest was theirs; but once out in the open they were subject to all the evil influences of the villagers, particularly witchcraft. The theory was that they would be made independent because they would no longer have to rely on the villagers for plantains, rice, peanuts, or other village foods. In point of fact the only thing for which the pygmies might really be considered to depend on their neighbours was metal for their knife and spear blades. And this was the one thing the model plan for liberation did not teach them to produce or work for themselves.

It may have been coincidence, or it may have been the result of all this talk of pygmy villages and the way that elsewhere the forest was being cut down by modern machinery to build new roads, to open up new mines, or to cultivate more coffee: but within a few days what had been a happy-go-lucky camp became silent and depressed. The unhappy pygmies argued that the more the forest was cut down for roads and mines and foreign plantations, the more it would have to be cut down to make more plantations to feed all the labourers. It was not a happy outlook for them. Then, as always happened when they were upset, they began to sing one evening. They sang and sang, and even those who had been sleeping came out to join in.

During a pause between songs, while everyone was silent, we heard quite distinctly the long, drawn-out whistling cry that the pygmies say comes from the chameleon. There was a shout of delight and the singing immediately started again, but this time to a quite different tempo. The rhythm was given out by the sharp clapping of two short sticks, which Amabosu had quickly cut from a nearby tree. They rang clear and true. The cry of the chameleon and the use of the *ngbengbe* sticks meant only one thing; there was honey nearby. Within a few days there was not a pygmy left in the village. Of all the many foods in the forest, honey calls the pygmies more strongly than any. At all times of the year there is game to be caught, different kinds of mushrooms and fruits and nuts and berries to be gathered, but the forest only gives its people honey during two happy months. It gives with even greater abundance, however, and the honey season is a season for abandoned merrymaking, where even the cares of hunting become unimportant as each individual, young or old, man or woman, sets off every morning in the quest for this precious gift. Above all, it is a season for rejoicing, for dancing and singing.

It is also a season when hunting groups split up into even smaller units, each going its own way, claiming certain honey trees for its own by marking them with a vine tied around the base, or some other sign that they know will be respected by all other pygmies. Only the villagers would be capable of breaking such trust. And so it is also a time when the pygmies tell stories, for the benefit of the villagers, of the dreadful and dangerous spirits that they have to fight in order to get the honey. Naturally enough, then, it is a time when the villagers prefer to leave the pygmies, and the forest, well alone.

Cephu was already off on his own, slyly camping near the valley that divides the territory of the Epulu pygmies from that of the next group to the west. He had relatives in both groups, and he felt able to claim trees in both territories, depending on where he thought he would do best. Those that were left split into two bands, one setting off with Ekianga, far away to the north of Kaweki's fishing camp. There they hoped to combine good hunting with the honey quest. A few of them made a permanent half-way camp so that they could act as a convenient relay station between Ekianga and the village, using the surplus meat to obtain luxuries that would help make the season even more festive.

Masisi, however, refused to have anything to do with this plan. The honey season was no time to be bothered with hunting, he said, and in any case he was fed up with village food and village tobacco and dirty village water. He wanted to forget about the village and eat nothing but forest food. He decided on an area that was little frequented, and there was a great deal of dissension amongst those who otherwise would have followed him without question. But Masisi was adamant and said he was going anyway; whether anyone else came or not did not worry him. His cousin, Manyalibo, was in a quandary. Morally he was bound to support his relative, but he was tempted by Ekianga's idea of getting the best of both worlds. Visions of both meat and honey won, and he announced that he was going with Ekianga.

Masisi, with the true pygmy flair for drama, said that Manyalibo was a disgrace, and that as everyone else obviously wanted to behave like villagers he would go off on his own. In a loud dramatic voice he commanded his wife to pack her belongings. Within an hour or so they set off, followed by their two children, Akobo and Ageranga. There was some discussion after they had left, and Njobo's sister, Teningbenge, and her son, Maipe, decided to follow Masisi. With them went Masamba, Njobo's wife. Njobo himself was still in Bunia with his son, looking after the boy who was being treated for severe tuberculosis of the bone. The next to follow them was old Asofalinda, who said that she wanted nothing to do with her brother, Ekianga. If Ekianga intended to waste his honey season hunting, he could do so without her. This somewhat counterbalanced Manyalibo's desertion of Masisi, and everyone thought it a great idea. She packed her few things together and set off, still denouncing Ekianga as a fool, carrying her crippled little daughter on her side, her bundle slung over her back. Lizabeti was a pathetic child whose hip had been diseased since birth. As a result she had never walked, but had spent her ten years of life crawling around on her buttocks, dragging one withered leg after her, using the other leg and her two hands to swing her body along as she vainly tried to follow other children in their games. Mambunia, Masisi's brother, himself with a leg twisted by infantile paralysis, went limping after Asofalinda, shouting that he would come and help her carry Lizabeti. As he left, clutching a spear in one hand, he told me that Masisi would be very glad if I came to their camp, I would find it less noisy than Ekianga's.

I asked Kenge what he wanted to do, and without hesitation he said he was going to follow Masisi. Maipe was his half-brother, and a very good friend of his, but he said that his main reason was that he was tired of noise. He knew that Masisi's camp would be a quiet one, with lots and lots of singing and dancing. To the pygmy, noise is synonymous with trouble, with dispute and discontent; music is never referred to as noise, but rather as 'rejoicing'. I felt that Kenge was right, and we decided to follow the next day.

It was, I knew, to be my last trek into the forest with the people I had come to think of as such close friends, and even though I knew I still had another two months with them I was sad. It was partly because of this, and partly because, like Kenge, I was so profoundly content to be back in our own part of the forest; and partly because Kenge also knew that this would be our last journey together, that I found the forest more beautiful than it had ever been. Kenge said it was just because it was the honey season, but as we passed the plantation and entered once more into the shade of the great, friendly old trees, he added: 'This is the *real* world . . . this is a good world, our forest.' He, too, was glad to be home.

We walked slowly, enjoying every minute. We passed through the plantation that Njobo and Masisi had begun to cultivate more than a year and a half earlier and found it a ruin, happily abandoned for more worthwhile things. We skirted Apa Muhoko because Cephu had camped there recently, Kenge said, and it would be full of jiggers. In the deserted remains of Apa Kadi-ketu we stopped for a rest, sitting on a log that had once resounded to the hammering of ivory bark-cloth beaters. I thought of the great *molimo* songs that had been sung there, and wondered if I would hear them again before I left. Beyond Apa Kadi-ketu we followed the path to Apa Lelo, almost to the spot where I had once seen a leopard crouching up in the trees above me, and there we turned west, following an animal trail that was narrow and tangled. It led through hilly country completely new to me, and finally brought us down to a swampy region, dark and dank. A stream flowed through the middle, fed by innumerable tiny rivulets from the hills all around. We picked our way across, balancing on fallen tree trunks and jumping from log to log; and as we began to climb up a steep hill on the far side I heard the tell-tale sound of bark-cloth being pounded into shape, the cheerful shouting of women as they called to each other while they finished the building of their huts, and an occasional outburst of song, sung for no other reason than that this was the honey season, the season to be happy, the season to rejoice. Kenge gave a loud whoop of pleasure and darted among the trees like a young antelope, leaping and bounding out of sight. When I reached the clearing he was already sitting down, his face buried in a sticky *mongongo* leaf, dripping with rich brown honey. He looked up and laughed. The next moment he thrust the leaf at me, and I was soon as sticky, and almost as happy, as he.

The camp was called Apa Toangbe, Toangbe being a special kind of honey tree. It was built half-way up a fairly high hill, on a patch of ground that was not quite level. Masisi's house was at the highest point of the tiny clearing, with Asofalinda to one side and Maipe's mother to the other. Masamba had built a house that virtually joined on to that of Masisi. Mine was half-way up the slope on the other side, facing across to Maipe. It was all ready when we arrived, and the only one to be completely finished. Masisi proudly led me to it and inspected it with me. The bed inside he had built himself, carefully refraining from turning up the ends sharply the way pygmies seem to think it should be done, so that it becomes the shape of a hammock without any of its back-saving resilience.

It was not thirty feet from one side of the camp to the other, and the forest closed in a protective wall all around us. Kenge and I set off to see what the surrounding country was like, and soon we were far beyond the sounds of any voices other than our own. Everywhere we went we saw mushrooms and nuts and berries, and we collected all we could carry to bring back for dinner. The hill was ringed with streams, each one fresh and clear until it reached the swamp. On the way back, having made a complete circuit, we were just nearing the camp when I noticed a huge old tree, one of those with great buttress roots joining the trunk some fifteen feet above the ground. It was completely dead, standing up white and naked amongst all the greenery; its lifelessness seeming all the more pitiable in the midst of such luxuriance. Most of its branches had already come crashing down, only one or two jagged stumps remained jutting out from the main body. Kenge gave it a critical look, then peered in the direction of the camp a few yards below. 'It should just miss our house!' he said, quite seriously.

As a matter of fact it did. One night we were awakened by a sharp, cracking sound, followed by a loud groan. This was repeated, and after a while a number of pygmies, each with a flaming brand in his hands, stood out in the clearing, listening. 'It's not going to fall yet,' they said, and went back to sleep. Kenge had not even bothered to get up, but I could not sleep. I must have lain there for a couple of hours, listening to the cracks as sharp as a rifle going off, and the constant groaning. Then the groaning got louder and more continuous, and in a flash Kenge was up and outside, yelling at me to follow. Just as I cleared the doorway the old tree came down, slowly and reluctantly, clutching at all the lesser trees that stood in its way. They were living, and it was dead, but it smashed its way on downwards to the ground. The tremendous din of its collapse was followed by a light pattering of broken branches and twigs as they followed after it, and then silence. Kenge said, 'You see, it did miss our house.' So it had, but by no more than six feet.

For the first day or two I noticed the children playing in a vine swing that they had rigged up for themselves, and I watched the crippled Lizabeti as she eyed them longingly. Several times, when the other children had gone scampering off into the forest I saw her pull herself over to the swing, slowly and painfully, and hoist herself almost upright by catching on to the vine with her hands, using what strength she had left in her good leg, fast withering in sympathy with the other and through lack of use. It suddenly occurred to me to ask Kenge why pygmies never made crutches, and he said that they either got better with the use of a stick, or else they just crawled until they died. What good is anyone who can't walk? he asked. I went to Asofalinda and asked her if she would let me try and teach Lizabeti to walk. The old woman was in one of her most truculent moods and severely said no. The white doctor had tried once before and Lizabeti had only got worse. Besides, the one leg was completely disjointed now, she said, and the other would be equally useless soon. I tried to explain that I was not going to try and cure the girl, but just to help her get around more easily. Asofalinda asked me what crutches were like, but then she immediately became severe again and said that everyone knew it was impossible to walk on one leg.

So Kenge and I went off into the forest and cut ourselves two pairs of crutches, carefully choosing sticks where there was a wide fork that would not pinch under the arm. We cut another pair the right size for Lizabeti. We padded the forks with leaves, and then I showed Kenge how to use them, holding one leg doubled up behind me. He was delighted, and was soon swinging himself unsteadily over the ground. When I thought that we had both had enough practice we returned to the camp and went up to Asofalinda, showing her the sticks and explaining them to her. We then performed, circling the camp twice without mishap. By this time the children were shrieking encouragement, and parents came running to see what the noise was all about. We performed once again for Masisi, and when Asofalinda still hesitated Masisi simply went into her hut and grabbed the terrified Lizabeti and carried her outside. He told Asofalinda that this was not medicine, it was just common sense, and she must not refuse. Did she want her child to die like other cripples who had been unable to fend for themselves?

Reluctantly she agreed, but Lizabeti, as soon as I reached for her, burst into tears and dragged herself back into the furthest corner of her mother's hut. Masisi started to go in after her again, but Kenge called to him to leave her alone, and handed his crutches and those we had made for Lizabeti to the eagerly waiting children. Soon it was the most popular game in the camp. Even Masisi and Maipe tried their hand at 'walking on one leg'. I saw Lizabeti's tear-stained face peering from the door of her hut to watch.

The next morning Asofalinda came over to me and said that Lizabeti wanted to try this new game, but if she fell once she would never be allowed to try again. Not a soul left the camp that morning, they all wanted to see what would happen. Between us Kenge and I lifted the girl, who was bravely trying not to cry, and put the crutches under her arm. We showed her how to hold them with her hands at the side, and how to take the weight off her good leg and swing it backwards and forwards. The greatest difficulty was getting her to trust that leg enough to move her crutches on in front of her. But little by little, moving one crutch at a time, she got more confidence. I went in front to catch her if she fell, and Kenge held her lightly around the waist from behind. Lizabeti began to walk. And then, as she realized that Kenge was not even holding her, that she was actually walking upright, like the other children she had watched from the ground for all her life, her face became contorted with fear and she stumbled. Kenge caught her just in time. He swung her high into the air with a triumphant shout to drown her wails of terror, and all the children joined in laughing and chanting their congratulations.

We had almost completed a circle of the camp, and I looked over to see if Asofalinda had seen Lizabeti beginning to fall. I was just in time to catch her burying her face in her hands, her whole body shaking as she sobbed her heart out—probably for the first time in her wonderful, hard-bitten old life.

The days that followed were full of happiness for everyone. Once Lizabeti got over her fright I made her practise regularly, to give her more strength in her good leg. She was soon able to stand up on her own, using the crutches for a lever. Not once did she fall. And the other children cut themselves crutches so that they could join in the fun, and together they found a new world of delight. When I was not coaxing Lizabeti to use her crutches more and more, someone else was, for everyone was anxious to see her fly over the ground just like Kenge, who by now was a master acrobat on his crutches, from which he refused to be parted.

In other respects, however, the camp became less pleasant, at any rate for me. For the bees began to come in droves after their stolen honey. All day long they buzzed angrily around, so that we kept smoky fires in our huts night and day. Almost every hour someone returned from a secret forage, with leaf bundles tied to his belt, dripping the sticky liquid down his legs. Sometimes it was too liquified to be eaten, and then the whole bundle was simply dipped into a bowl of clear forest water, making a sweet tasting drink. But far more popular was the whole comb which could be eaten, grubs, larvae, bees and all. If it was very hard then it was softened over the fire first, and this made the grubs squirm more actively so that the honey worked its own way down your throat. The first time I tried this I confess I had a hard time, and Kenge laughed for hours, miming the way I closed my eyes and let him put it into my mouth, irrationally not wanting to touch the dreadful mess with my fingers. It was, however, the best-tasting honey of all, despite the insects, and it was certainly replete with very active vitamins.

Apuma was yet another kind, but it came later in the season. It was full of grubs too, but they were all dead, and the honey had fermented. It was made by a different kind of bee, and tasted like a bitter liqueur; if eaten in any quantity it could be highly intoxicating. But the pygmy needs no stimulant in the honey season. He is drunk with the forest, with its beauty and abundance, and with the love it showers on its people. Every night that tiny camp resounded to songs of joy and praise, accompanied by the ringing of the *ngbengbe* sticks as they were clapped together in complex cross-rhythms by boys and girls alike. And every night Masisi told his family, which had steadily grown ever since the beginning, stories about the past. And every night we went to bed content, knowing that the morrow would be even better; for each day we discovered fresh growths of mushrooms, trees full of different kinds of nuts, and each night the chameleon gave its long, sad cry, to tell us that on the morrow we would have more honey.

One night in particular will always live for me, because that night I think I learned just how far we civilized human beings have drifted from reality. The moon was full, so the dancing had gone on for longer than usual. Just before going to sleep I was standing outside my hut when I heard a curious noise from the nearby children's *bopi*. This surprised me, because at night time the pygmies generally never set foot outside the main camp. I wandered over to see what it was.

There, in the tiny clearing splashed with silver, was the sophisticated Kenge, clad in bark-cloth, adorned with leaves, with a flower stuck in his hair. He was all alone, dancing around and singing softly to himself as he gazed up at the tree-tops.

Now Kenge was the biggest flirt for miles, so after watching a while I came into the clearing and asked, jokingly, why he was dancing alone. He stopped, turned slowly around and looked at me as though I was the biggest fool he had ever seen, and he was plainly surprised by my stupidity.

'But I'm *not* dancing alone,' he said, 'I am dancing with the forest, dancing with the moon.' Then with the utmost unconcern he ignored me, and continued his dance of love and life.

As the weeks went by, however, a note of dissatisfaction stirred in the camp. Masisi began to complain that the others, including Manyalibo, were way off to the north getting meat, yet they had sent him, their own brother, none. And why did they not come and join him? I expect Ekianga was saying the same thing about Masisi in the south, with all his honey. But one day I decided to go and see for myself. With Kenge I set off early in the morning. After several hours of rapid trekking we came to a deserted camp. The fires still smoked, and it had obviously not been long abandoned. We scavenged around and found two spears and a bow and arrows that had been left behind, presumably in the hurry to get away before we arrived. There were also several baskets and pots. Armed with the more worthwhile trophies we returned to Apa Toangbe. Masisi was wild with fury at this deliberate insult, and told Kenge and myself to go off again the next day and find them, and bring him back some meat.

We found them a considerable distance away. They were thoroughly surprised, having thought we would have been shaken off and discouraged. They hurriedly tried to conceal the great amount of dried meat that had been accumulating, but they could not hide it all. So they criticized Kenge for having been so mistrustful as to sneak up on them silently from the side, instead of using the main trail. Kenge replied that in our experience we found that main trails only led to deserted camps, and he then delivered Masisi's comments in full. The comments were pithy, and Kenge added his own embellishments. He dwelled mainly on family loyalties, and there was no doubt that Manyalibo was wrong because he had even taken Masisi's hunting net, promising to send meat, and had never sent any. Njobo's net was there too, yet his sister, Maipe's mother, had received none of the meat it had trapped. Ekianga himself had sent no meat to Asofalinda.

I had expected a spirited defence, but Manyalibo slowly came out of his hut, where he was hiding, and said grandly that we could take all his meat back to Masisi; he had had enough, and anyway he had really been saving it for his 'brother'. He added that as for joining Masisi, there was not much sense doing that now, as everyone would probably be moving camp soon, why didn't we all meet up later somewhere near Apa Kadi-ketu?

Kenge and I returned, laden down with meat, and with protestations of friendship. Much the same thing happened when Masisi thought he would try similar tactics with Cephu. This time I went with Kelemoke, who had come to stay with us because he said he found Cephu's camp far too noisy. It was not much more than three hours away, and it was plain that Cephu had settled in well with his friends and relatives, because he was almost in the middle of a large and industrious camp. Everyone seemed to be eating well, and Cephu offered us food, which was unusual for him. He did not part with any meat, but he had already heard of our haul at Ekianga's camp, and asked us why we had brought *him* none. I told him that Masisi thought it would be a good thing if he joined us and the others, and that he would get plenty of meat then. Cephu said he would think about it, but he was offended that we had brought him nothing.

Masisi pondered over these things for several days, and our evenings became interspersed with brief arguments between him and Mambunia and Kelemoke as to what they should do. Everyone was beginning to be hungry for meat again, and although Ekianga and Manyalibo had promised to send our nets back, they never came. We did not have enough in our own camp to make a successful hunt. In the end Masisi made up his mind to beat the others to it and get to Apa Kadi-ketu before they did. By this time we had been joined by a number of bachelors from both the other camps, also by a third group of distant in-laws of Masisi, all of whom had learned that we were getting a great deal of honey.

On our way to Apa Kadi-ketu we heard news that brought matters to a head. We met Madyadya coming hot-foot from the direction of the village, to warn Ekianga that a group of foreign pygmies, from the other side of the Epulu, had invaded our territory and were stealing all our honey to the east. Masisi immediately sent his son with Madyadya to tell Manyalibo to forget the quarrel about the nets, and to come and join him at once at Apa Kadi-ketu so that we could all make war on the other pygmies together. He also sent Kelemoke with the same message to Cephu. We pressed on and built our camp in a hum of excitement.

Kenge saw that I was taking this all very seriously, and he took me on one side. 'Old Masisi's head is unsteady,' he said. 'It has worms in it. He will only fight with words. Every year those pygmies come into our land and we go into theirs. There is plenty of food, so long as we do not meet there is no fighting. If we do meet, then those who are not in their own land run away and leave behind whatever they have stolen. That is the only way we ever fight—we are not villagers.'

But it was a wonderful fight of words, particularly when Ekianga and the others joined us. There was so much disagreement that Masisi said he was not ready to live in the same camp with them yet, and he removed himself to another camp about half a mile away. Cephu, still not sure which way the wind was blowing, camped between the two, and to one side, making a perfect triangle. And for a whole week messengers were being sent backwards and forwards. Eventually the invaders, presumably having had their fill of honey, retreated across the Epulu and said that never again would they enter such inhospitable territory.

But everyone was far too busy scouring the countryside and marking out new honey trees to be seriously bothered. It was strange to see how this familiar old camp took new shape. Any vestiges of old huts were cleared away, except for the remains of the one where I had lived, and that of Ekianga, both of which had been made better and bigger. I did not take my old hut, but moved to the far end, where Cephu had been, near to Moke. Ekianga and all his three wives, now on good terms with each other, were to the other side of me. Masisi, still breathing fire and fury about the treachery of his relatives, came to visit us from time to time, otherwise he kept strictly to his own camp.

And so, after taking advantage of the one season they know, a season which allows them to forget about the cares of communal hunting, the need for constant co-operation, the camp was beginning to come together again. Every year this splitting up takes place, and every year the forest draws its people back together. Whatever frictions had existed were soon healed, a new year was due to begin. Before this camp was over, Masisi would have found some excuse to move a little further north, Ekianga was already talking of shifting a few hundred yards south, and Cephu would make up his mind to jump this way or that. Everyone knew it, and within a week of arriving at Apa Kadi-ketu they were talking about where they would build their next camp, and how it would be one camp again . . . until the next honey season.

But I would never know that next camp, so I made the most of my last home in the forest, sharing it as always with a good and constant friend. Maliamo set up house in Masisi's camp, and sometimes Kenge stayed there, sometimes in the main camp. At all times the forest was full of the song and dance of life. The games special to the honey season filled the daylight hours and at dusk there was the dance of the honey-bees. Tremendous vine swings were slung from the highest branches, and on these the youths practised their acrobatics, leaping and somersaulting in the air. When a vine broke it was used for interminable games of tug-of-war, men against women. But there was a conspicuous lack of gallantry in this particular game, for if either side saw that it was losing ground it sent a saboteur into the other ranks. It is difficult to concentrate on pulling at a vine when someone is flirting with you—as pygmies do.

But best of all was the dance. Once again the men and the women divided, and while the men pretended to be honey gatherers, dancing in a long, curling line through the camp, looking up with exaggerated gestures as if searching for some sign of bees, the women danced in another long line through the trees at the edge of the camp, pretending that they were the bees. The two lines gradually came closer and closer together, the women singing in a soft, rhythmic buzz, buzz, buzz, while the men pretended to hear, but still not to see them. Then the women grabbed up burning logs of wood and attacked the men, tapping the logs on their heads so that a shower of sparks fell over them, stinging them like the sting of the honey-bee. Then they all gathered up the embers, and where some of the younger men had been building an elaborate hearth of special woods and special leaves, moistened to just the right extent, they lit the great honey-fire. There was no flame, but dense clouds of smoke billowed upwards. Men blew on their honey-whistles, women clapped their hands, and everyone burst into the song of magic that would travel with the smoke and call the bees to come and make more honey.

On such occasions I found that Masisi and Cephu made their appearance more and more often, paving the way for a reunion at the next camp. When the time came for me to leave all animosities had been forgotten.

One day Kenge had gone off on some unknown mission and I decided to make a final pilgrimage to Apa Lelo, where the cycle had begun. This day of all days I wanted to be by myself, but there is something about the forest, not exactly threatening, but challenging, that dares you to travel alone. I had to refuse several pygmies who wanted to come with me, and once I was well away I was glad. I knew what the challenge was: for to be alone was as though you were daring to look on the face of the great God of the Forest himself, so overpowering was the goodness and beauty of the world all around. Every trembling leaf, every weathered stone, every cry of an animal or chirp of a cricket tells you that the forest is alive with some presence.

As I splashed my way through the last stream, and climbed to the top of the rise above Apa Lelo, the sounds of the forest grew louder. This was a friendly sign, but somehow it made me feel even more lonely. Then I came down into the old camp, the huts rotted and collapsed, the clearing overgrown. As I wandered about, tracing where this hut had been and who had lived there, kicking idly with my feet in the hopes of finding some familiar old object, it all began to live again. I wandered down to the Lelo, as calm and placid and beautiful as ever, and I thought of the *molimo*. As I looked into the waters where the trumpet had once been hidden, it was as if I heard its voice in the distance. And when I bade my farewell to that very special river and returned to the old camp, I saw it as it really was in the timelessness of the forest world; the huts strong and green, the fires burning brightly, and the air heavy with the scent of cooking. Lizabeti was running about with the other children, swinging happily in her crutches; the men helped each other to examine and repair their nets, while the women cooked and gossiped. Masisi was shouting at someone, and Manyalibo was trying to pacify him in his usual gruff, good-natured way. Old Moke sat back in his chair and puffed at his pipe, gazing up at the tree-tops so very high above. I heard him murmur once more, 'You will see things you have never seen before . . . You will understand why we are called "People of the Forest" . . . When the forest dies, we shall die.' And for the last time I heard the chorus of that great song of praise: 'If Darkness *is*, Darkness is Good'.

The camp and the people disappeared, leaving only a shimmering haze over Apa Lelo; but the song remained, for the song is the soul of the people and the soul of the forest. The *molimo* trumpet, now infinitely wistful and far away, took up the song and the forest echoed it with its myriad magic sounds. It echoes on and on, and it will still be there when the sounds of our short lives are silenced; until, perhaps like us, it comes to rest in the deepest distance of some other world beyond, the dream world that is so real to the people of the forest.

Index

251

relationship with pygmies, 25, 127–9, 131, 145, 157–67, 209–13, 216–19; and adoption of customs, 149, 167, 180–81, 182, 196, 201, 202, 204, *and see nkumbi* ceremony; ambiguity of, 157–8; blood-brotherhood with, 197; disputes with, 160–61, 182, 193–5; friendship with, 161–6; and gambling, 127–8; and gifts, 128–9, 157, 213; gulf between, 204; and honey, 237; and lack of control, 210–11, 212–13; and law, 211–12; and pygmy 'liberation', 233–6; and marriage customs, 182, 185, 186, 190–95; 'ownership' of, 192, 210–13; patronage of, 124–5, 145, 149, 154, 156, 157–8; trade with, 25, 35, 145, 146, 157; and wars, 221
and spirits, *see under* spirits
and witchcraft, 24, 38, 46, 50, 87, 205–6, 216
women, *see* women, negro
vision, and distance, 227–8

wailing, 44–5, 46, 48, 49; and crime, 103–4
walking, correct way of, 56, 75–6
washing, 55, 58, 60, 70; and death, 149; and *elima*, 180–81; of *molimo* trumpet, 73–4, 84
weddings, *see* marriage
witchcraft: BaLese and, 24, 216; BaNgwana and, 154–5, 162–3,

206; and illness, 46, 205, 206–7; pygmies' disbelief in, 46, 51, 205, 207; villagers and, 24, 38, 46, 50, 87, 205–6
witch-doctor, 155, 198–9
women, negro: gift of, 128–31, 211; male attitude to, 186; and menstruation, 167–9
pygmy: and beauty, 68, 136–7, 149, 170; clothing of, 22; and childlessness, 112, 137; decoration of, 68, 89, 136, 176–7, 190–91; and decoration of clothing, 120–21; and *elima* celebration, 133, 141, 149, 167, 176–7, *and see elima* celebration; exclusion of from *molimo*, 76, 77, 78–9, 83, 140; and hunting, 93, 94, 95–6, 97, 115, 140; and hut-building, 62, 63–4, 65, 67, 122–3; importance of, 186; lack of discrimination against, 140, 186; and menstruation, 169, *and see elima* celebration; and mourning feast, 149; occupations of, 89, 90, 120–22, 131, 152; participation of in *molimo*, 134–41; relationship with men, 102, 130–31, 185, 186; and travel, 53, 58, 62; *and see elima* celebration, marriage

youths: behaviour of, 55–6; and *molimo*, 29, 76–7, 81–3, 134–5; and the village, 143; *and see* bachelors, boys